ALTADENA LIBRARY DISTRICT
3 9270 00376 2949

Mother Nature Is Trying to KILL You

A Lively Tour
Through the Dark Side
of the Natural World

Dan Riskin, PhD

A TOUCHSTONE BOOK
Published by Simon & Schuster
New York London Toronto Sydney New Delhi

Touchstone
A Division of Simon & Schuster, Inc.
1230 Avenue of the Americas
New York, NY 10020

First Touchstone hardcover edition March 2014

Interior design by Aline C. Pace
Jacket design and illustration by Shasti O'Leary Soudant
Jacket photographs © iStockphoto

Manufactured in the United States of America

10 9 8 7 6 5 4 3 2 1

Library of Congress Cataloging-in-Publication Data
Riskin, Dan.
Mother Nature is trying to kill you : a lively journey through the dark side of the natural world / Dan Riskin, PhD.
pages cm
"A Touchstone book."
Includes bibliographical references.
1. Predation (Biology). 2. Predatory animals. 3. Predatory insects. 4. Parasites. I. Title.
QL758 .R57 2014
591.5'3
2013032852

ISBN 978-1-4767-0754-9
ISBN 978-1-4767-0756-3 (ebook)

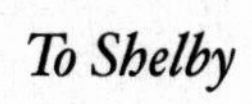

To Shelby

CONTENTS

Mother Nature Is Trying to KILL You

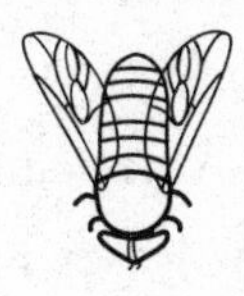

INTRODUCTION

Georgia on My Mind

It wasn't that it hurt all that much. In fact, most of the time I didn't feel anything at all. But once in a while I'd get a sharp pinch, almost like someone was pressing the edge of a spoon into the top of my head. I didn't feel sick, and I knew this thing wasn't going to do any permanent damage, but it was driving me nuts: a small, white, pasty maggot with rings of black bristles around its body was lodged firmly into the top of my head. For all my squeezing, I hadn't been able to pop it out. It was only a few millimeters long now, but it was eating my flesh, and I knew it would get much, much bigger in the next few weeks. For the life of me, I had no idea how I was going to get it out.

I'd gotten it by accident (obviously) a few weeks prior, working in Belize as part of a team studying bats there. Bats are my thing. I've always loved bats and by studying them I've been able to travel all over the world: I've been to Australia, New Zealand, Madagascar, South Africa, Costa Rica, all over North America, and even to the Amazon rainforest of Ecuador, all just to see bats. The excursion to Belize was early in my career, before most of those other trips, and looking back, I see now that I still had a lot to learn about the natural world.

Belize is a small Central American country, about the size of New Jersey, but whereas New Jersey has nine species of bats, Belize has fifty. That's why I was there. It was 1998, and I was in the first year of graduate school. Our team wanted to find out where some of those Belizean bats hide during the day, and it was a great opportunity for me to get some experience identifying species I'd never seen. After reading countless books and papers about bats, this was my chance to get out and meet some of them.

The place where we worked in Belize is called Lamanai. It's famous for its ancient Mayan ruins, but we were there for the lush rainforest those ruins are buried beneath. The wildlife in Lamanai blew my mind. There were big, colorful toucans, six-foot crocodiles, brightly colored venomous snakes, howler monkeys, about a zillion kinds of gorgeous insects, and of course, bats! Fishing bats, vampire bats, yellow-shouldered fruit bats, sac-winged bats, sword-nosed bats, frog-eating bats . . . Trust me, if you've got even a passing curiosity about bats, Belize is heaven.*

It was a two-week trip, and we were pretty much constantly

---

* I participated in a couple of studies that documented which bats live at Lamanai (Fenton et al. 2000, 2001), but even more species have been identified there since we wrote those papers.

By the way, you'll see footnotes like this one throughout this book. They're intended to provide added context to the stories you're reading. Anytime you want to dig even deeper, just follow the little numbered endnotes to the Notes section at the back of the book.

bushwhacking, so it's impossible to guess exactly when I got the botfly. Each night we'd set up nets to catch bats; then we'd put radio transmitters on some of them. A transmitter is about the size of a coffee bean with a five-inch antenna hanging from it. You glue it to the fur on a bat's back, then let the bat go. The transmitter emits a beeping noise on a specific radio frequency, and you can tune in with a radio receiver to find out where the bat is hiding.

Once a radio-tagged bat is released, you can only pick up its signal if you're near the bat, or if you have a clear line of sight between the hidden bat and the antenna in your hand. So each day, we'd climb to the top of an ancient Mayan ruin and scan for bats above the forest canopy. Once we got a signal, a beeping from the north, for example, we went out with a machete and cut a trail northward until we picked up the signal again from within the forest. Then we hacked away through branches and vines, following the signal as it got stronger, until we could finally see our bat roosting on a branch, or at least figure out which tree hole our bat was hiding in. It was tricky because we needed to cut a path in front of ourselves with machetes, but we had to be quiet so as not to scare the bats off as we approached. It was also tricky because most of the vegetation we were hacking through was full of nasty things like scorpions, thorny acacia plants covered in ants, and venomous snakes. But, man, I have never had so much fun in my life. By the end of the trip, I was feeling pure bliss. I was sunburnt and covered in bug bites, but I'd found a career path that excited me, and I was starting to imagine filling the rest of my life with adventures like this one. Unfortunately, in all the excitement I just failed to notice that one of my insect bites had a maggot in it.

The human botfly looks like a normal housefly, but it is far more stomach-churning than the thing that keeps landing on your salad.[1] An adult female botfly zips around in the rainforest, then catches a mosquito in the air, lays an egg on her abdomen, and lets her go. Later, the mosquito bites a mammal (a monkey, a jaguar, or a bat biologist, for example), and while that mosquito feeds, the botfly's egg falls off the mosquito and onto that mammal. The egg hatches into a maggot, which makes its way into the hole created by the mosquito and then settles in to feed and grow. The maggot starts out just a few millimeters long but grows steadily, until after a month and a half or so, when the maggot—now about an inch long—eats its way back out of the skin, falls to the ground, molts into an adult fly, and then flies away.

It's actually a pretty smart strategy by the botfly. The adult is big enough that if I'd seen one land on me, I'd have slapped it silly. But the adult never came anywhere near me. It used a mosquito as its courier service, so I didn't even notice. In fact, I didn't realize anything was wrong until I was home a few weeks later.

The maggot was firmly lodged at the top of the back of my head, just on the right side. At first it was like any mosquito bite, but a bump started forming around the bite, and that bump was steadily growing. That spot is a very hard place to look at closely on your own head. I tried with two mirrors and a flashlight, but I just couldn't get a good look—it was too hard to part my hair and squeeze with all that stuff in my hands. All I could see was a swollen red area about the size of a dime. In the center of the red zone was a tiny hole, and sometimes other people would tell me they could see a little white snorkel come poking out of that hole. That was how I knew for sure what I had, because that snorkel is how botfly maggots get their oxygen.

I'd never had a botfly before and didn't really know how to remove one properly, so I just kept trying to squeeze it like a zit.* This did not impress my friends in the slightest. They made me wear a hat, and if I so much as scratched my face, they'd immediately make me go wash my hands. I tried to explain to them that they couldn't catch a botfly from me, but that didn't seem to matter. As far as they were concerned, I had cooties.

After a couple of days, they felt comfortable enough to start making fun of me, but they still wouldn't come very close. Their jokes were marvelous. They told me the whole botfly thing was just in my head. They told me they knew it was tough but that I shouldn't let the botfly get under my skin. And they also wanted to know if I could claim my botfly as a dependent on my taxes. They even gave her a name.

Georgia.

So it went, for about a week. I went about my business, with Georgia on my mind, squeezing to try to get her out now and then but mostly just hoping she'd somehow go away. Botflies are hard to squeeze out because they have sharp backward-pointing bristles around their bodies that brace them firmly in the flesh. If they're lodged somewhere like your arm or chest, you can squeeze enough skin to get behind them and pop them out of that breathing hole, but other parts of the body don't let you get underneath it. On my head, I just couldn't get the leverage I needed, so I'd end up squeezing Georgia on the sides, and then she would just hurt more. I hoped that once she got bigger, I'd be able to get my

* Some people recommend you lure botflies out by covering their air holes with pork fat. It's not the smell of food that draws them out, though, it's the lack of air. That's why surgeons in one study were able to draw a botfly out by covering its breathing hole with Vaseline, which blocked the maggot's access to air (Liebert and Madden 2004).

fingers underneath her to squeeze her out, but that strategy never seemed to work.

Maybe if I'd waited longer, I would have been able to dislodge her with my fingers, but I reached my emotional breaking point long before that. About two weeks after coming home from Belize, I snapped. I was in my car, driving to the grocery store, when that spoon-edge-into-my-head pain suddenly started up again. It was in that moment that I immediately decided that Georgia was coming out. I hoped a doctor would agree to do it, because I didn't want to have to ask a friend. I drove right past the grocery store, made a right, and headed to the hospital.

It's not easy to gross out a nurse. They see vomit and feces all the time. But it turns out (at least in Canada), that telling a nurse you have a maggot lodged in your head will do the trick. That made me weirdly popular in the ER, so despite a pretty busy waiting room, it didn't take long for me to see a doctor.

He came in, looked at my chart, then looked me right in the eye and told me he didn't know what a botfly was. I have this weird paranoia that doctors don't take me seriously to start with, so I tried to sound as knowledgeable as possible. I told him all about the breathing hole and the bristles and the life cycle with the mosquito, but it was hurting as I talked, so I was kind of hunched over and wincing while I explained. That made me worry he would think I was nuts, and that made me talk even faster. It wasn't going well.

He looked me straight in the eye again and said nothing for several seconds. Then he took a deep, disappointed breath and asked to have a look at my head.

He put on some latex gloves as I lay chest-down on the table. I put my chin on my crossed arms and he started poking around at the top of my head with his fingers.

"I don't see anything."

"Right here." I poked it.

"That? There's nothing there," he said. "It maybe looks like an ingrown hair."

"No, no. It's a botfly. See the hole?"

He kept poking. Then he sighed again. He was right on the fence. Either he believed me and I'd be getting rid of this thing right now, or he was going to send me home, and I was going to have to talk one of my friends into cutting my head open.

Finally, after a long pause, he asked me how deep he would need to cut.

"I don't know. Cut as deep as you want. There's skull between you and my brain, so you can't possibly do too much damage." I was only sort of joking.

I heard him take another breath, followed by a long pause. Then I felt the prick of a needle, freezing the area. Soon I felt a little tug, and then blood started pouring down my forehead in the front, and down onto my neck in the back.

He passed me a towel and I put it across my forehead. In just a few more seconds this thing would be out of me. I couldn't wait.

But then the doctor said, "I still don't see anything."

If the doctor couldn't get it out, my plan B wasn't going to work either. Now I had a big cut in my head. Were my friends going to have to dig around it? Reopen it? Maybe I *was* imagining things. What if it wasn't a botfly? Then what was it?

I didn't respond. He kept tugging or cutting (they felt the same at that stage) for what seemed like several minutes, and then he made a quiet, surprised gasping noise.

I didn't feel anything at all. I didn't want to turn my head to see what he was doing because I had the towel on my head just right to soak up all the blood.

"Did you get it?"

He rolled his chair around in front of me so I could see him, and presented me with a small urine-sample container full of alcohol. Near its surface, Georgia floated lifelessly.

I was finally free.

Today Georgia sits in that same urine-sample container, on a shelf next to my desk at work. She's only a few millimeters long, disappointingly small compared to the other botflies I've seen on the Internet. Apparently, a lot of people don't know about botflies, so when they get them, they just let the mystery sore grow for six weeks or so until, to their total surprise, a maggot writhes out of it. Georgia was removed long before that stage, but even though she's smaller than some of the others, she's mine. I earned her. I'm proud of her. Ask me if I've ever felt like I was "at one with nature," and I can hold her up and show her to you. She's like a medal.

I went to Belize as a scientist, to study how biological organisms live in nature, and instead I experienced nature *as* a biological organism. That really influenced how I perceive the natural world. To me, nature's not just a pretty photograph of a rainforest. It's constantly changing, twisting in a dynamic life-and-death drama driven entirely by a battle for energy—energy that flows from host to parasite, from prey to predator, and from rotting carcass to scavenger, in a never-ending battle among all creatures to persist and pass on DNA. Belize reminded me that I'm part of that epic system. We all are. My botfly fed on me to get energy, and I spent energy trying to get rid of it. That's what nature is—a place where animals selfishly try to survive and make babies by getting the upper hand on one another.

When you think about nature that way, it becomes strange that lately everyone and their dog seems to be telling us we should be living *more* naturally. We're surrounded by advice on diet, exercise, medicine, and lifestyle, all championing some version of a "natural" way to live. Since humans evolved as part of nature, we're told that escape from our modern problems is as simple as getting back to our roots, by moving/eating/behaving the way humans did millennia ago.

The main flaw with that kind of advice, of course, is that it ignores the basic fact that a few thousand years ago people typically only lived into their thirties. There are predators, parasites, and decomposers everywhere, ready to break us down for food at the first opportunity. Living in a modern, Westernized society, we've done such a good job of keeping most of those threats at bay that those "natural" advice givers can act as though things like botflies, rattlesnakes, and malaria don't even exist. But believe me, they do.

If anything, Mother Nature is trying to kill you.

My perspective on nature is a little bit different from the way nature is typically portrayed, but that's often because advertisers and marketers are using the word *nature* to sell you something. In their world, nature's a benevolent bounty of well-being that can always make you healthier and would never, ever hurt you. It's all honey and no stingers. Nature gives us fruits, vegetables, and shade-grown coffee beans, but not the mold around the bathtub or ants or tapeworms. Advertisers would like us to think those things aren't really part of the green world. Instead, they're somehow invaders of it.

Think of a typical shampoo commercial with scenes of soft meadows and waterfall-fed pools, where beautiful models frolic and no two strands of hair ever cling together. There are plants and maybe butterflies, or even horses, but there are never any hornets, scorpions, or leeches. Icky animals hurt sales, so the image of nature has to be left half-finished.

With that utopian version of pseudonature established, companies can then boost sales by making you associate their products with that friendly world. I was at the grocery store recently and saw a cleaning solution that said "nontoxic," "organic," and "green" on the label. It also said, right on the front of the package, that it kills mold, mildew, and bacteria. One wipe of that cleaner across the kitchen counter can snuff out millions of lives, but they call the product eco-friendly, as though somehow bacteria don't count as part of "eco" (whatever that is). I'm not arguing against using green products—in fact, I encourage it—but look at the irony of the labeling. I mean, the organic apples and oranges you buy aren't grown in nature, they're grown on farms. How are they more natural than the mold in your home that came in through the window?

Because magazine editors, TV executives, and advertisers don't like to acknowledge the dark side of nature, we don't see the whole picture and we end up with a warped perspective.

When I was an undergrad, I worked in a restaurant with someone who once told me that she would never do any drugs made in a lab but that "if nature made it, it's okay to take it." When she said that, I remember thinking it *sounded* like it might be logical, but that turned out to be one of those statements you just have to think about carefully for a minute or two before you realize it makes absolutely no sense whatsoever. I think she was talking about marijuana and magic mushrooms, but heroin, crack cocaine, and cigarettes all start as plants too. Whether you're talking about recreational drugs, food, or anything else, pretending that nature is 100 percent benign requires either ignorance or willful disbelief.

Let me put this another way: From the window seat of an airplane, the skyline of New York City stands enormous, majestic, and silent. But when you visit Midtown, it's a totally different ball of wax. What looked like faceless rectangular buildings from the air turn out to be beautifully decorated, with stone and brick patterns that make each of them unique. Standing on the ground, you can hear honking cars, people shouting into their phones as they walk, and trucks rattling by as they roar over filthy potholes. You can smell garbage and urine, but you can also smell hot dogs and pretzels. You can get a bite to eat, go see a show, take the subway, or just sit down on a bench to take it all in.

New York's skyline is beautiful, but it leaves a lot out. In fact, I'd argue it leaves out all the best parts of New York.

To see nature as an endless bounty of health is as incomplete as describing New York by her skyline alone. It's true that nature has produced many, many things that are good for us, but nature has also created jellyfish, fire ants, and cyanide. We can celebrate the healthy parts, but when we take a look at the creepier side of nature too, the whole picture gets richer. In fact, I'd take it one step further: I think the disgusting, immoral, and violent side of nature, the side that the grocery stores and shampoo commercials leave out, contains its most awe-inspiring and beautiful parts.

Walking along a trail in Belize looking at birds through binoculars was like seeing the skyline of New York, but my botfly experience took me down to street level. It changed my relationship with nature by getting me outside my comfort zone. And a couple of years later I embarked on another up-close-and-personal experience with the natural world—one that would last much longer than my adventure with Georgia. I thought getting a maggot lodged in my head was life-changing, but it turns out that's nothing compared to having a baby.

In August 2011, my wife and I welcomed our first child, Sam, into the world. Since that day, I see everything differently. I used to weigh myself every single day just because I loved collecting the data, but now I'm simply not that interested. I used to fill my calendar with trips all over the world to see as many bats as I could, but now I would rather stay home with Sam than spend a month catching bats in Africa without him. It's not just that I care more about this little kid than I've ever cared about anything before (and I do), but Sam's birth also changed how I feel about everything else—including myself. You've probably heard someone else say this before, and if you're a parent you might have even said it yourself. It changes you. And you're changed because you feel something that you don't feel from any other experience. Most people would call that feeling "goodness" or "pure love"—but in truth, I'm not so sure that's really what it is.

Here's why I hesitate to use that language: you can't say that botflies are "evil," since they're just doing their best to survive and ensure their own DNA's survival to the next generation. It makes no sense to talk about a botfly's behavior in terms of "good" or "evil," and the same holds true for any animal in nature.

Since the feelings I have about Sam very clearly come from my own biological drive to survive and protect my own DNA, I don't see why the concepts of good or evil should apply there either. What feels like fatherly love is really just my body's way of protecting its own DNA in the next generation. If the botfly's behavior isn't "evil," why should I call my feelings about Sam "pure" or "good"? Yes, I experience something that feels wonderful, but that's just what my DNA tricks my brain into believing. Is there anything *real* behind those illusions?

I wrote this book in part because I'm passionate about the ugly, heinous natural world and want to take you on a tour to see that. But I also wrote this book as part of a personal journey of my own. I need to figure out whether the love I feel for my son is real.

So let's begin. In order to avoid getting lost, I've organized our tour around a list of evils in the world that were outlined by the Catholic Church about 1,400 years ago as vices that could bring out the very worst in humans. You've heard of them, I'm sure. They're called the seven deadly sins: greed, lust, sloth, gluttony, envy, wrath, and pride. Of course, there's nothing biological about sins—the whole idea of morality is a human one. But I think it's going to be a fun challenge to ask whether Mother Nature might in fact perform each of those seven sins with even more flair than humans do. Along the way, either the roots of fatherly love will be found in the jungle of selfishness, or I'll know for certain that the love I feel for my son doesn't really exist.

So let's start searching.

1

GREED

Lions Don't Kill Zebras. Zebras Kill Zebras.

A biology prof, an engineering prof, and one of their students walk into a bar. Just as the three of them are taking their first sips of beer, a lion walks out of the bathroom, looks at them, then licks its lips.

The biology prof slowly rises from his barstool, then stands frozen with terror. "In the last twenty-five years there have been more than a thousand lion attacks on humans, two-thirds of which were fatal. I don't like our odds at all."

The engineering prof looks at the lion, then the door, then starts doing some calculations. As she stands up, she says, "I don't like our odds either. Based on its size, I'd estimate the top speed of this lion at around forty miles an hour. Olympic sprinters can go twenty-three miles an hour. Even if we could run half that fast, which I doubt, the lion would get to the exit well before we would. We're screwed!"

The student looks at the lion, looks at his professors, pauses, and then pulls a phone out of his pocket to take a photo of the lion for Twitter.

"What are you doing?" asks the biology prof. "Can't you see that there's no way for you to outrun this lion?"

"Absolutely," says the student, finally getting up from his seat. "But I don't have to outrun the lion. I just have to run faster than one of you."*

\\\\\\

That joke's been around forever, but it doesn't get old because it's basically true. A predator is only going to catch you if it doesn't catch someone else first. Every year in South Africa, plenty of zebras get killed by lions, but they don't die randomly.[1] Some zebras are better at avoiding lions than other zebras are, because they're faster, more attentive, healthier, or maybe just luckier. Lions do the killing, but *which* zebras live or die is determined by the zebras. They're like the three people in the bar: so long as a zebra can run faster than other zebras, it has a better chance of survival than the slower ones do.

---

* The profs have their facts straight. More than a thousand people were attacked by lions in Tanzania alone between 1990 and 2007 (Kushnir et al. 2010). Lions often hunt men guarding their cattle or children playing outside, but sometimes lions even pull people out of their beds. This inconvenient aspect of lion behavior throws a major wrench in conservation efforts, since the people who live in lion habitat have a very real incentive to get rid of them (Packer et al. 2005). Also, the top running speed of the African lion is 37 miles per hour (Garland 1983). Usain Bolt won gold at the London Olympics by running the 100-meter dash in 9.58 seconds. That gives him an average speed of 23.4 mph over the whole sprint.

The first of the seven deadly sins on our tour of the living world is greed, and by that I mean an all-consuming focus on yourself. I'm talking about Ebenezer Scrooge–ness. In *A Christmas Carol*, Scrooge didn't want to have to pay his clerk, Bob Cratchit, for Christmas Day if Cratchit wasn't coming to work. Scrooge also refused to give even a shred of his considerable wealth to the poor. All Scrooge cared about was himself. Period. That's how animals make decisions—except while Scrooge was greedy about money, animals are greedy about passing on their genes. Animals will screw one another over without any hesitation, just so long as it helps get copies of their own DNA into the next generation.

Some of the most aggressive actions animals take in the name of selfishness are carried out against members of their own species. They do this because often the most threatening animals in the world are those that are most like you. Members of your own species want to live in the same place as you, they want to eat the same food as you, they want to have their babies in the same place as you, and they want to feed their babies the same kind of food you want to feed your babies. If you ever have to fight a member of your own species, there's a good chance you'll be injured. To pass on its DNA, an animal has to make sure it doesn't get the short end of the stick compared to those other members of its own species, and the best way to do that is to be as greedy and selfish as possible.[2] This might sound jaded, but it's absolutely true, and a recent study demonstrated this point beautifully using sheep as an example.

It's been known for a long time that when a predator like a wolf approaches a flock of sheep, the sheep in the flock will cluster closer together. People have assumed that this was because the sheep were all running away from the same predator and therefore just ended up in the same location, but that's not what happens at all. Sheep are smarter and far more conniving than people give them credit for.

Researchers fitted an Australian flock of sheep with GPS collars so that they could see how individual sheep moved in response to an approaching predator (in this case, a herding dog). When the dog approached, the whole group closed in on itself as expected, but not because the sheep were all running away from the predator in unison. Instead, the GPS data showed that individual sheep reacted to the dog by moving toward the center of the flock—even if that wasn't directly away from the dog. In other words, the direction chosen by a sheep wasn't determined by the location of the predator, it was determined by the locations of the other sheep. Using the same strategy as the student in the bar, a sheep can survive without having to outrun its predator. It just has to put other sheep between itself and the source of danger. A sheep can live, just so long as someone else dies instead.[3]

There's a similar story with emperor penguins in Antarctica. During the impossibly cold and dark winters there, males huddle together for more than three months while each incubates a single egg. Males don't eat anything over that time, and by spring, each penguin has lost more than a third of his body weight. Penguins are extremely well insulated, but they still leak a little heat to the world around them. Because all the other penguins are giving off heat the same way, the best possible place for a penguin to stand is right in the middle of the huddle, with other penguins leaking heat on every side of him.*

* Male emperor penguins go without food for 105 to 115 days and drop from an average of

This works better than I would have ever imagined. A penguin in the middle of a huddle experiences temperatures like those on a tropical island, between 68°F and 99°F, while those near the edges are exposed to some of the coldest temperatures on Earth.[4] On the surface, this huddling behavior looks like a great example of teamwork. In fact, it was part of a very popular documentary released in 2005 called *March of the Penguins*, narrated by Morgan Freeman, who said,

> *Though they can be aggressive during the rest of the year, at this time the males are totally docile—a united and cooperative team. They brace against the storm by merging their thousand bodies into a single mass. They will take turns, each of them getting to spend some time near the center of their huddle, where it's warmer.*

That's a nice way to describe it, Morgan Freeman, but they're not really a team. None of the penguins is keeping track of who's taken his turn in the middle or at the edge, and none of them willingly leaves that warm middle for a turn on the cold edge. Instead, all the penguins act selfishly, and the morphing huddle is what happens as a result.

A detailed analysis based on hours of video showed that the overall movement of the group is exactly what you would expect if penguins follow a simple, selfish set of rules: If you have penguins on all sides, don't move; if you're on the edge, try to shove your way back into the middle. You *could* call that "taking turns," as Morgan Freeman does, but when you think about it,

82 lb. to an average of 53 lb. Penguins are only able to incubate eggs successfully when they huddle with other penguins (Ancel et al. 1997).

it's really what would happen if every penguin was just trying to stay as warm as possible.* The decrease in aggression among male penguins that the movie mentions makes sense from that selfish perspective too; for a huddling penguin it would be counterproductive to kill the heat source next to you. But for other animals in other situations, killing thy neighbor is the best possible strategy, and when that's the case, they don't hesitate at all.

A male zebra who comes across a baby zebra that isn't his own will kick and bite it until it dies.† By doing that, he reduces the competition that his own offspring will face in life. Similarly, when a male lion displaces a rival and takes over his group of females, the first thing he does is kill the babies in his newly inherited pride so that he can start making babies of his own.‡ Killing babies means more food for his (soon-to-be-conceived) babies and also has the added bonus that the females will come into heat sooner than they would if they were still nursing their young. Killing babies might at first seem unthinkable, but for many animals it's just another way to look after their own selfish interests.

Remarkably, babies are even sometimes killed by their own siblings. Take snowy owls, for example. These large white birds are found in the Arctic, where they eat lemmings—small rodents that go through population explosions every once in a while. That

* That study showed that the ability of the group to stay warm depended on multiple factors, including the speed of the wind, the number of penguins huddling, and the shape of the huddle as seen from above. Overall, though, an individual penguin's movements could be best described by the rules I've mentioned. It's impossible to know what a penguin is thinking out there, but they act as though they are trying to keep themselves warm, with no regard for other penguins (Waters et al. 2012).

† This behavior is a big problem for zoos with captive zebras (Pluháček and Bardoš 2000).

‡ A lion cub normally has a 56 percent chance of living to the age of nine months. However, if a new male takes over the pride, the cub's likelihood of survival plummets to 14 percent (Packer 2000).

means that in some years lemmings are very hard to find while in other years you can't swing a dead lemming without hitting a lemming. The problem is that those population explosions are unpredictable, so the owls don't know ahead of time how many eggs to lay. In some years there are enough lemmings for a pair of owls to feed a whole nest of owlets, but in other years there's hardly anything around. Fortunately, the owls don't have to guess. They can just let their owlets fight it out themselves.

First, mama owl lays one egg. Then she waits two days and lays another. And so forth, until a total of five to ten eggs are laid. When the first owlet hatches, it gets all the food it needs from the parents and starts growing instantly. By the time the second egg hatches, that first owlet is quite big, putting the new, younger sibling at a real disadvantage. Whenever mom arrives at the nest to barf up lemmings for the babies, that first chick pushes its weaker sibling out of the way and gets as much vomit as it wants. Only after number 1 can't eat any more does the younger owlet get to eat. As the subsequent eggs hatch, that system continues. Only after number 2 has had enough will number 3 get any, then 4, and so forth. This system's brilliant, because it means in a lean year, instead of having six chicks so hungry that none of them has a chance of survival (which is what would happen if all the eggs hatched at once), the parents end up with one or two very healthy chicks. In a great year they end up with more healthy chicks. Owls can use this system to raise exactly the right number of chicks every year, no matter what the lemming populations are, and all it requires is that the parents let the babies enforce their age-based pecking order.[5]

To be fair, the snowy owls don't actually kill their brothers and sisters, they just starve them by leaving them no food. But that's not the case with Verreaux's eagles of Africa. I don't know if you ever suffered abuse at the hand of your older sibling, but it can't possibly compare to what Verreaux's eagles go through. A female of that species always lays two eggs, but only one survives. The older sibling gets about a three-day head start on its little brother or sister. Then, as soon as the second bird hatches, the physical abuse begins. It's relentless. In one study, a researcher watched an older sibling peck its younger one 1,569 times, until it died at the age of three days. This abuse happens every single time. Out of two hundred observed nests, the younger sibling has been observed surviving in just one.[6]

It's not just birds either. Baby sand tiger sharks eat their own brothers and sisters before any of them are even born. At first, they develop within egg capsules inside their mother's uterus, where each embryo gets its energy for growth from a yolk, just as in a chicken's egg. But eventually that yolk runs out, and unfortunately, that happens before the shark is ready to be born. So the first shark to reach that stage of development (the oldest) swims around the uterus, eating other egg capsules and the siblings inside them. In fact, it doesn't just eat the other egg capsules at random but actually seeks out the egg sac of the next-largest sibling and eats that one first. That way, there's no chance of a strong competitor in the uterus later on.[7]

The fact that animals will even kill their own brothers and sisters is a good reminder that animals aren't born to be nice. They don't try to preserve their own species, and they certainly don't care about the other species of the ecosystem living around them. Instead, each animal looks after itself. An ecosystem is what emerges when every animal in nature follows Scrooge's philosophy. (In fact, Scrooge comes off looking like a pretty good guy when you compare him to a Verreaux's eagle.)

But of course, at the end of *A Christmas Carol* (spoiler alert), Scrooge changes his ways. The ghosts of Christmas past, present, and future collaborate to scare the crap out of him and convince him that no matter what the short-term benefits of greed, it's a bad strategy in the long term. The ghosts tell Scrooge that if he doesn't give up on his greed, he'll become a ghost like them instead of going to heaven. That scares Scrooge enough that he immediately changes his attitude, making Christmas better for everyone, including his faithful clerk, Bob Cratchit.*

So what about all these selfish animals? Won't their greed ever bite them in the ass? What would ghosts from the past, present, and future say to them? The truth is, greed usually goes unpunished, but in some instances that shortsighted strategy can lead to catastrophe in the longer term. And there's no more wonderful example of that than the giant carnivorous mice of Gough Island.[8]

Gough Island is a cold and rainy rock in the middle of the South Atlantic Ocean, but millions upon millions of seabirds breed there. After all, many of these birds feed on fish out in the middle of the ocean, so they don't have a lot of places available to lay their eggs. The island is only thirty-five square miles in area—that's less than half the size of Lincoln, Nebraska—but it's been called "the most important seabird island in the world" because more than twenty species breed there, some of which live nowhere else on Earth.[9] There aren't any trees, so the birds lay

* It might occur to you that Scrooge technically remained selfish, even after the ghosts visited him. Once the ghosts pointed out to him that his money wouldn't help him after he died, Scrooge decided to put his efforts into securing a comfortable afterlife. He never stopped looking after his own interests, he just shifted his focus from getting richer to getting into heaven. The fact that he suddenly became kind to everyone else was just a by-product of his selfishness.

their eggs on the ground, and their chicks live there until they're old enough to fly. That nesting strategy worked wonderfully for millennia because there were no predators on the island. But everything changed about two hundred years ago, when humans visiting the island by boat accidentally dropped off some common house mice.

Mice are usually vegetarians, but every once in a while, if food is scarce, a mouse will scavenge on the carcass of a dead animal. It's rare, but it does happen.

When those mice first got to the island, around 1810, some of them might have scavenged on the corpses of whatever dead birds happened to be lying around. But at some point, a few of the mice on Gough Island figured out that you don't have to wait for a seabird chick to die before you can eat it. Having lived on an island free of predators for so long, the seabirds had no strategy for defending themselves from land predators. The mice realized they could pretty much just start eating seabird chicks alive.

Biologists visiting the island have watched this happen. A chick will sit defenselessly in its nest, while a mouse walks up and starts chewing on its torso. That attracts other mice, and soon a whole group of ten to fifteen mice will work away on the skin and connective tissue of a chick, battling one another for the meat, and eventually leaving a hole in the bird's back up to two inches in diameter—that's big enough that you can clearly see the internal organs through it. Not surprisingly, the bird soon dies from those injuries. This makes no difference to the mice, which just keep on eating until nothing remains but bones.*

* Some thoroughly depressing footage of mice eating a baby albatross alive is available as "supplementary data" in a paper by Wanless et al. (2007).

These aren't small chicks either. Most weigh at least a pound, and researchers have even seen mice take out a juvenile albatross weighing more than twenty pounds! Because of that relatively large prey size, it can take the mice several days to kill a chick. To me that sounds like absolute torture, plain and simple, and seabird populations are plummeting on Gough Island as a result. But of course, the selfish mice don't give a rat's ass about that.

It sounds like paradise for the mice, finding themselves on an island with an unlimited food source, but it's actually been pretty tough for them too. Not long after they arrived, the island became totally overrun with carnivorous mice, so it got harder and harder for individual mice to compete with one another. It was no longer enough to be able to digest meat. To survive, a mouse now had to be able to fight off mobs of other meat-eating mice. In general, larger mice had an advantage over the smaller ones in those fights, so larger mice got more food, made more babies, and the mice of Gough Island started getting bigger and bigger with each passing generation. Today, the average size of a mouse there is around thirty-five grams. That's *two to three times* what a normal house mouse weighs. That increase in size happened over the course of just two hundred years. However, the greed that made them into superstars may well have also painted these mice into an evolutionary corner.

Because the mice are eating all the chicks, seabird populations are plummeting, and it's pretty clear that the birds will be gone before too long.[10] That's when things will suddenly get very, very bad for the giant meat-eating mice of Gough Island. At first they'll probably just eat one another,* but ultimately the island

* Even normal mice have been known to cannibalize one another. Mouse mothers fre-

will no longer be a place where carnivores can survive. Perhaps a few mice will revert to vegetarianism, but their large body sizes, and other body changes associated with carnivory, might make it hard for them to get enough calories on a vegetarian diet. If the mice had done a little planning, they could have lived merrily alongside the seabirds, eating plants and chick carcasses for centuries. But they didn't, and couldn't, plan ahead. There's no long-term planning built into the process of natural selection. That's where selfishness, even though it's good for an individual in the moment, can be bad for a species as a whole.*

Since the beginnings of life on Earth, the story of the Gough Island mice has played out countless times, but with other species in other places at other times. Animals move to a new environment and thrive, but their presence there changes things until those same animals can't survive there anymore. And as humans, we know a thing or two about changing an environment once we get there.

When humans arrived in Australia, fifty thousand years ago, they began systematically wiping out all big animals.[11] All nineteen mammal species there weighing more than two hundred pounds vanished soon after the arrival of humans. More than half of all species weighing between twenty and two hundred pounds disappeared as well. We know and love the kangaroos, koalas, and wombats of Australia today, but if humans hadn't been such a plague on that continent, we'd also know a four-thousand-pound, hippo-sized wombat that stood six feet tall at the shoulder, a five-

quently eat their own babies (Rowe et al. 1964), and adult-adult cannibalism happens too (Resendes et al. 2009).

* When animals evolve themselves into a corner like that, it's called selection-driven self-extinction (Parvinen and Dieckmann 2013).

hundred-pound kangaroo so heavy it couldn't hop, and a lion with a pouch like a kangaroo's. Some of those animals were hunted directly by humans, but humans also cleared trees and burned large areas for agriculture, and that would have wiped out many herbivorous animals by robbing them of the plants they depended on.[12] The Australia of today seems pristine and precious to us now, but its plants and animals are a small subset of what the first human settlers found there. I'll say it again, because I think it's hard to comprehend the first time you read it: more than half of Australia's big, charismatic animals are gone.

North America suffered a similar fate when humans arrived there, between ten and twenty thousand years ago.[13] Mammoths, saber-toothed cats, and giant ground sloths all vanished within a few thousand years of human arrival, leaving the subset of animals we know there today. The pattern was similar to that in Australia, with everything over two thousand pounds disappearing, along with more than half of the species between seventy and two thousand pounds. Bison and grizzly bears and elk are pretty cool, but sometimes when you're walking through Jasper National Park, in Canada, you can't help but think it would be nice to have mammoths and camels around too.*

Animals have also vanished from the islands of the South Pacific as humans have colonized those islands over the last four thousand years or so. Most people have heard of the dodo bird, which went extinct on Mauritius, in the Indian Ocean, in the 1600s, but large, flightless birds were once quite common

* Interestingly, the North American extinction seems to have been caused by human activity in concert with climate changes that were happening at the same time (Faith 2011; Prescott et al. 2012). Those changes in climate were not caused by humans, though, unlike those happening in the world today.

on many islands of the South Pacific. Islands like Samoa, Fiji, Hawaii, New Zealand, and Rapa Nui (Easter Island) all suffered extinctions upon the arrival of humans.[14] In total, humans have wiped out roughly a thousand large bird species (including the dodo) from those islands, and extinctions of other animals, including bats, snakes, and many other animals, continue to this day.

Humans have a consistent track record upon arrival to new lands that makes us look an awful lot like the mice of Gough Island. But since humans really are just animals, it seems fair to say that human short-term thinking and greed are just our animal instincts shining through. In the pleasantries of modern society we may conceal our true colors, but fundamentally we're all just giant mice. Proof of that lies in brief moments of panic, when humans act on their instincts. If you have any doubt of that, just look at the animalistic ways humans behave during shipwrecks.

When a ship sinks at sea, everyone knows the rule: Men are supposed to help "women and children first," because otherwise women and children wouldn't have a fair chance. It helps to be big and strong when you're fleeing a sinking ship and getting to a life raft. You have to move quickly through corridors and up stairwells, perhaps with debris and other people in the way. There may also be swimming involved, in high ocean waves or frigid seawater. Since men are typically stronger and more aggressive compared to women (and certainly compared to children), this is where men would otherwise have an unfair advantage. So the "women and children first" rule is there to make sure men don't act selfishly.*

---

* To be clear, there are many women who would outcompete the average man in a compe-

Here's the natural experiment: If people are fundamentally generous, men should always help women in maritime disasters, so women should have higher rates of survival in those scenarios than men do. If people are selfish, men should survive in greater proportions than women. So what happens?

In a broad study of eighteen different maritime disasters spanning the 1850s to the 2010s, researchers found that the likelihood of survival for a woman was about half that of a man.[15] That suggests that even if men do save some of the women sometimes, they tend to save themselves in greater numbers. Similarly, crew members survive better than their passengers do, because they know where the life rafts are and how to work them. This isn't just a problem from olden times either. In the 2011 sinking of the Russian ship MV *Bulgaria*, 60 percent of men survived but only 27 percent of women. In 2012, when the cruise ship *Costa Concordia* ran aground off the coast of Tuscany, its captain infamously got himself to shore well before the evacuation had been completed. Thirty-two people died.

Of course, everybody's favorite maritime disaster is the RMS *Titanic*, which ran into an iceberg one chilly North Atlantic night in 1912 and then slowly sank, killing more than 1,500 people. It's a tragedy, no doubt, but for some reason it's a tragedy people can't seem to get enough of. Well, if you're looking for something about the *Titanic* to feel good about, try this: the RMS *Titanic* is one of only two disasters of the eighteen in that study where people actually followed the "women and children first" rule. The survival rate for women in that disaster was 70 percent, compared

tition of that kind (my wife, Shelby, can run faster and farther than I can, for example), but on average, men have the upper hand. In a random sample of men and women, you'd expect the fastest 10 percent to be made up of more men than women.

to just 20 percent for men. So what made the *Titanic* special? Why did men follow the rules then, but not in other disasters?

The answer, it seems, may lie with an unlikely hero, the *Titanic*'s captain, Edward Smith. Shortly after the collision with the iceberg, Captain Smith gave an order to his crew that women and children should be saved first. His crew enforced that rule aggressively throughout the evacuation, effectively preventing selfish men from saving themselves. There are even written accounts of crew members shooting at men who tried to get on board the lifeboats that night. The crew enforced the "women and children first" rule so that selfish male passengers couldn't act on their instincts.

It's tempting to look at historical records like those and say, "Yeah, but *I'd* act differently," but I challenge you to ask yourself if that's really true. Drowning would be a horrible way to die. Whether you're a man or a woman, would you honestly give up your seat on a life raft for someone you didn't know? If you saw an empty seat on a lifeboat, would you really just walk away from it so a total stranger could live?

I suppose no one can really know how they'd act until they are in that position (heaven forbid), but I'd like to think that if I were in a shipwreck I'd do everything I could to save as many women and children as I could—men too, for that matter. But I think it's fair to say that the second I felt like my own life was in danger, I'd get my ass on that lifeboat. Wouldn't you? I want to be a good person and all, but fundamentally, I have a lot more invested in my own well-being than in that of some random person I don't know.

Here's where things shift, though; here's where being a dad changes everything.

If I imagine my son with me on that hypothetical sinking

ship, not only would I let other people die so Sam could live, I would kill them myself, if I had to. In fact, I would even sacrifice my own life if it meant Sam would survive.

I think almost any parent would do the same for their child, and animals make similar decisions all the time. A songbird will try to save its eggs by dive-bombing an approaching cat, even though that puts the bird itself in danger. If a baby water buffalo is tackled by a lion, its mother will use her horns to gore that lion, despite the fact that the lion might turn to attack her instead. In life-or-death situations, parents instinctively put their own lives on the line for their young, because that's often the greediest way to behave. After all, they're not being greedy about surviving, they're being greedy about passing on their DNA.

In fact, let me go one step further, and this is a bit of a mind bender, so bear with me: the animal isn't the one protecting the DNA at all. The DNA protects itself. The animal's body is an elaborate robot, made of meat and bones, built by the DNA to protect itself. That's really the most accurate way to describe it. It's a sort of meat robot.

The animals of the world, ourselves included, are kind of like those giant AT-AT walkers from *The Empire Strikes Back*. From a distance they looked like independent creatures, but they were really just four-legged vehicles, built and piloted by people, for a war against other people. You can think of our bodies the same way. Each of us is built by DNA molecules—DNA molecules that are at war with other DNA molecules. Animals' bodies, our bodies, are just the meat robots of DNA molecules in battle.

There's an important difference between AT-ATs and animals, though. During a battle, AT-ATs are piloted by the people inside them. DNA can't do that because DNA really has no idea what's going on out there. Instead, the DNA builds its animal to function autonomously. It's built with an innate set of urges and instincts, like the urge to breathe, the urge to avoid pain, and the urge to have sex. Those instincts keep the animal alive and coerce the animal to reproduce. As a result, the DNA survives and is passed on to the next generation, so it can make the next generation of AT-AT-like meat robots.

When you look over the rail of a high balcony and feel your very soul pulling you back away from the edge, that's your DNA's programming at work. When a new crush takes over your whole world—and when holding hands with that person makes you feel like your heart's going to explode—that's the DNA pushing your buttons. To call them "urges" almost doesn't give them enough credit. They're instincts. They're the things that matter most to us—life, love, and sex—and they're all just programs written into our meat robot bodies by the DNA inside us.

I was first introduced to this way of thinking about twenty years ago by the book *The Selfish Gene* by Richard Dawkins, and ever since I read that book, I've seen myself differently.* For example, if I'm feeling stressed about a big talk I need to give or about a meeting with my boss at work, I always seem to be able to step outside the situation to see myself as a worked-up meat robot. When I do, I somehow let go of a bit of that stress. When I see myself as a programmed machine, the stakes seem much, much lower.

---

* *The Selfish Gene* (Dawkins 1976) is among the very best introductions to natural selection out there. The "meat robot" theme I use throughout this book is largely inspired by Dawkins's shift in perspective from the individual to the DNA molecule. His book is definitely worth a read for that reason alone. As a bonus, it is the book where the term *meme* was first coined, so it's interesting to read for that reason too.

That way of thinking has thrown me for a loop since Sam was born, though. I don't like to think of my parenting skills as DNA-written urges. It's depressing to think I love my son simply because that's what my DNA tells me to do, because that means it wasn't really my decision to love Sam at all. It means I'm a zombie, caring about Sam because my DNA wants the copy of itself within Sam to survive. That's just sad.

It especially bothers me to think of Sam and me as meat robots when I'm spending time with him. He's a smiling, innocent, intelligent person, growing and learning and living right in front of me, but I keep overthinking things and looking at both of us as robots. Playing with Sam, reading him stories, chasing him, making him laugh—sometimes I'm with him, but I'm just not totally present because I know too well what's really going on. Sam's carrying 50 percent of my DNA, and my DNA is ensuring its own survival by making me think I love him. It's hard to play along enthusiastically when you realize you're a manipulated puppet.

And if that weren't sad enough, I know that when Sam kisses me and smiles at me, it's because his DNA is ensuring its own survival. We're just a pair of meat robots. Like all the other animals, Sam and I are just selfishly going about our lives, using one another as prescribed by the DNA within us. The love we feel for one another, and everything about our beautiful relationship, all boils down to selfish greed.

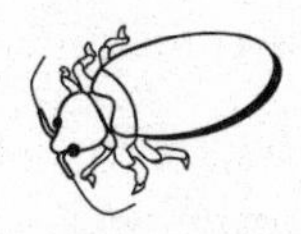

2

LUST

Meat Robots Screwing One Another . . . Over

Maybe you're worried I exaggerated the influence that DNA has on animals. Maybe you think that whole meat robot idea went a step too far. After all, animals have brains, right? Shouldn't that put them in a position of control over a twisty microscopic molecule? I know, I've made a bold claim, so I'll let the animal stories speak for themselves. You'll see that DNA really is in the driver's seat, and the best evidence of that comes from stories about sex.

Here's what makes sex so interesting: animals, including humans, face a trade-off between how long they can live and whether or not they reproduce. If an organism were really in control, it would make decisions that optimized its own survival, and then only have sex if that didn't shorten its lifespan. But that's exactly the opposite of what we see in nature. Time and time again, animals do things to pass on their DNA that cause their bodies harm. Animals will even die to have sex. Those animal decisions collectively build a convincing body of evidence that reproduction is even more important than survival; DNA really is in control, and the body really is just a meat robot. A sophisticated meat robot, but a meat robot nonetheless.

When I was a kid, I found it amazing that animals knew how to mate in the first place. It was no surprise to me that animals instinctively avoided danger, but for some reason I always had a hard time getting my head around the fact that animals instinctively know how to screw. Without training, animals know to thrust certain parts of their bodies against one another in very specific ways, so that sperm will come in contact with an egg. And it's not just the screwing that comes instinctively either: that horniness drives a male satin bowerbird of Australia to collect anything blue it can find to decorate the ground outside its home—berries, flowers, Bic pen caps, whatever—because blue things will attract a female.* Horniness makes male bighorn sheep smash their heads into one another's because that's how they can get a female.† All of that happens instinctively. To me, it's no surprise that animals are programmed to run away from predators, but I marvel at what they're programmed to do for sex.

* The male builds his "bower" out of twigs. It's a structure with two parallel walls, about four inches apart, with everything cleared off the ground around it. The male decorates that ground with blue objects, and the number of times females will mate with him depends largely on how many blue objects he can accumulate (Borgia 1985). One of the best places to find blue objects, of course, is the bower of another male bowerbird, so theft is a big part of competition among males (Wojcieszek et al. 2007).

† Head-ramming behavior is essential to rams that want to mate (Martin et al. 2013), but collisions tend to be quite bad for mammalian brains. Rams avoid concussions, though, by having heads that are built in a way that mechanically dissipates energy when they ram (Maity and Tekalur 2011).

When I was about ten, I had the same thought about humans. How do they know what to do in bed? I remember thinking that people must have to pass on the secret of how babies are made from generation to generation so that we don't go extinct. Otherwise, how on earth would anyone ever think to place a penis inside a vagina? Seriously? Who would ever think of that? I imagined scenarios where uneducated teenagers shipwrecked on islands would have no idea what to do. "Even if they did figure out to lie on top of each other," I thought, "what if they didn't know to take off their clothes? How would they figure that out?"

Once I hit puberty I realized my worry had been grossly misplaced. What I'd previously thought of as a very strange ritual between grown-ups had quickly become the only thing I could think about. In junior high my urge to have sex honestly felt about as strong as my urge to stay alive. My DNA had sent new instructions to my meat robot self in the form of raging hormones, and suddenly my whole soul was hell-bent on executing those directions. Girls were suddenly the only thing I cared about. I imagine that's probably how powerful those urges are for satin bowerbirds and bighorn sheep.

This chapter is about lust—that DNA-driven, overwhelming urge to make babies that impels animals to hurt one another, hurt themselves, and even sacrifice their own lives. I'll start by highlighting some of the most heinous sacrifices animals make for reproduction, from being eaten alive to having their clitorises torn open (you can decide which is worse). Along the way, I'll challenge you to think about your own assumptions about sex and nature. Is there such thing as a natural childbirth? Are homosexual relationships natural? Is rape natural? You'll quickly see that nature is so immoral, vulgar, and downright wicked, we can't possibly use nature's behaviors to set rules for ourselves.

Here we go.

\\\\\

No animal makes a more stunning sacrifice for sex than the male antechinus.[1] (If you want to say its name properly, make it rhyme with "acts of kindness.") Antechinuses are cute. They live in Australia and pretty much just look like big-eyed, furry mice, but they're really marsupials—pouched mammals more closely related to kangaroos and koalas than to mice. What makes antechinuses so unusual, though, is that every single year, they go through a mating season so intense that it kills every single male of the species.

Mating season is in August and lasts anywhere from three days to a couple of weeks. For a male antechinus that brief period is the only chance he'll ever get to pass on his genes, so he sacrifices everything to give it his best shot. Testosterone levels skyrocket to ten times normal; sperm production goes into overdrive. As mating season begins, males are so full of sperm that a little bit leaks out of them every time they pee. The males need all that sperm, though, because antechinus mating is like a marathon; copulation lasts for six to twelve hours.

Hours. (*I know*, right?)

That little mouse-sized ball of fur makes love like a tantric stallion. By taking his time, though, he defends that female from the advances of other males: mating season is so short that by staying attached to her for that long, he keeps other males off her for a significant fraction of the whole season. Also, that gives him time to put a significant volume of sperm deep into her reproductive tract. Whether his genes or someone else's will be passed on all comes down to his performance that week. He's not out to impress anyone (though I, for one, am very impressed). He's just doing it because he's got to pass on his DNA.

Whether or not that male knows *why* mating is so important is up for debate, but what is certain is that it's stressful for him. During mating season, male antechinuses experience off-the-chart levels of stress hormones. Those are the kinds of hormones released by people in the most stressful situations imaginable, for instance when people are displaced from their homes by civil war.* Those hormones, which come from the adrenal glands, tell the whole body how to allocate energy when times are tough: "Stop spending so much on the brain, kidneys, and immune system. Let's break down our fat reserves. Let's get our muscles ready for action!" Those hormones can help an animal survive hardships in the short term, but they can also start to hurt the body if the doses are too high for too long.

To mate with a female, a male antechinus must first win fights with other males, and those high stress levels help him get the extra boost of energy he needs to do that. But somewhere in the evolution of antechinuses, the levels of those hormones that males needed to win those fights got out of control. Males suffer kidney failure, ulcers, immune system breakdown, and a whole whack of other problems all because their stress hormone levels are just too high for their bodies to tolerate. By the end of mating season, every male is dead, having lived less than a year.

* I'm talking about cortisol. It's released in high doses by male antechinuses (Naylor et al. 2008), and by people with post-traumatic stress disorder, or PTSD (Steudte et al. 2011).

In experiments, researchers have tried cutting off the testes of male antechinuses, and those castrated males survive the mating season without any problems at all. That's the trade-off in a nutshell: if you want to mate, you're going to have to do things that shorten your life. If the male antechinus meat robots were in control, they would skip the stress of sex to live as long as possible, but it's their DNA behind the wheel, so the meat robots don't get to make that call.

A similar self-sacrifice for sex happens to a lot of male spiders, but they don't die from stress. They're killed and eaten by females during sex. It's called female cannibalism, and it happens in many different kinds of spiders. A female spider is almost always a much larger predator than a male, making him pretty much exactly the kind of animal she'd like to eat.* To have sex with her, a male has to put himself within striking distance—he has no choice about that. So males of some spider species bring food along as a "nuptial gift" when they want to mate. In those cases, the bigger the gift, the longer he gets to mate before she eats him, and the more of her eggs will be his.† That's a great system for a male when it's available, but most spider species don't exchange those gifts, leaving many male spiders out there no other option than to roll the dice and just hope the female isn't hungry.

* A lot of people assume male black widows are always killed by the female after sex, but males usually escape. In fact, male black widows can sometimes inseminate up to three females (Breene and Sweet 1985). The observation that females always kill males probably came from early experiments in closed containers that were so small the male couldn't escape, which isn't really how things would work in the wild (Vetter and Isbister 2008).

† An example of a spider species that uses nuptial gifts is the nursery web spider (Stålhandske 2001). Sometimes he'll play dead while he's holding his gift, and just when the female comes to take it, he'll spring into action and mate with her (Hansen et al. 2008).

One species of spider with female cannibalism is an orb-weaving spider called *Nephilengys*, but males of that species have evolved a survival strategy different from gift giving. Instead, they rip off their own penises mid-mating.

Well, truth be told, it's not really a penis. It's called a palp. And it's not just one palp; he has two of them. And they're on the sides of his head. Anyway, like any male spider, he mates with a female by inserting his palps into holes in her abdomen and pumping sperm into her. But partway through the mating process, the male *Nephilengys* spider pulls a quick one on her: he rips off his palps, leaves them sticking into her abdomen, and makes a break for it. He doesn't always escape—male survival rates are only about 25 percent—but even if she catches and eats him, his palps will keep on pumping sperm into her while she's distracted by the meal. It's not the cleanest strategy in the world, but it works.

If, by some great stroke of luck, the male does manage to escape, things get even more interesting. Now castrated, he waits just beyond the female's reach for any oncoming males with thoughts of romance. Should a male approach, the castrated male fights the intact intruder as aggressively as he can. Since he'll never be able to reproduce again, his DNA's only hope of being passed on lies with the female he just escaped. For his DNA's sake, the best strategy now is to prevent any other males from mating with her. That way the eggs she lays will carry his DNA, not someone else's.

It didn't take long for researchers looking at these spiders to notice that when fights happened between a castrated male and an intact male, the injured spider almost always won. This made them wonder whether there is something about having his genitals cut off that gives a spider a boost. So they did experiments chasing spiders around a table with paintbrushes until the spiders collapsed from exhaustion. Surprisingly, the researchers found that castrated males had 80 percent higher endurance than intact males. It seems that spiders with their palps intact have the instinct to give up early, since perhaps another female might be available elsewhere. The castrated male doesn't have other options, so he fights with everything he's got.[2]

Human males have their lives cut short by sex too. Most don't die from the stress of mating itself, and most aren't killed and eaten by their female sex partners, but they still pay a price. Specifically, it turns out that for humans, just having male sex hormones in the body shortens life expectancy. Evidence of this effect comes from some fascinating data on castrated men from eighteenth- and nineteenth-century Korea.[3] Back then, powerful men with harems would sometimes adopt eunuchs to work as guards or laborers. These men, who were castrated before puberty, were convenient employees to the emperors since they could do the manual labor but wouldn't try to have sex with women in the house. The eunuchs were perfectly healthy, but they had no testicles, and since testicles are the part of the body that produces testosterone, the eunuchs lived their otherwise normal lives without that hormone in their bodies.

Here's the evidence that human men live shorter lives than they would without sex: when you compare eunuchs to intact males of similar socioeconomic status in that same society, the eunuchs lived fifteen to nineteen years longer on average, typically into their seventies. Laborers with their testicles in place were living into their late fifties, whereas one of the eunuchs lived to be 109! This difference, coupled with the fact that females usually live longer than males anyway, strongly suggests that human males have their lifespans cut short by testosterone. Since you can't make sperm without testosterone, you could say that human males have their lives cut short by sex.

And as bad as it is for human males, they've got nothing on females of the species. Females can't pass on their DNA without going through pregnancy and childbirth, and although human childbirth is a beautiful and life-changing event, it's as dangerous as hell.

In North America, the odds that a fifteen-year-old girl will go on to die from a maternal cause (pregnancy, childbirth, or a failed abortion) is just one in 3,800, but those fantastic odds are due largely to the availability of modern medicine.[4] I'm talking about the basics, things like blood pressure monitoring before birth, sterile procedures during birth to avoid infections, antibiotics if infections do occur, and basic drugs to speed up clotting in case a mother bleeds too heavily after the baby's born. Those simple things improve a woman's odds of survival a great deal compared to what they are without medicine. As evidence, you can just look at places where those tools aren't available. The odds of a fifteen-year-old girl in sub-Saharan Africa dying from a pregnancy, childbirth, or a failed abortion at some point in her life is 1 in 150. At first, those better-than-99-percent odds might seem decent, but they're twenty-five times worse than the North American odds, and when you're talking about millions of women, those maternal deaths add up very quickly.*

* Maternal mortality per 100,000 births in a few Western countries: USA 21; Canada 12;

Take a moment to think about where you were twenty-four hours ago. Really do this. Stop reading, look at a clock, and figure it out. Think about exactly where you were and what you were doing.

Got it?

Since that moment, roughly eight hundred women have died painful and tragic deaths from pregnancy, childbirth, or a failed abortion, and more than 95 percent of those deaths would have been prevented by access to basic modern medicine. That happens every day, and those numbers are a concrete reminder that there's no Mother Nature looking after us.*

We're animals, and as with other animals, our bodies reflect a trade-off between survival and reproduction. Big-brained humans can compete better than small-brained ones, so the size of the average human baby's head has reached the very limit of what a mother can handle. If you imagine what childbirth must have been like even just a few generations ago, before painkillers, before antibiotics, nature starts to look pretty rough. Here in the developed world, that can be easy to forget sometimes.

When my wife, Shelby, and I found out that she was pregnant with Sam, we didn't even think about the possibility that she might die; we didn't really have to. If we'd bothered to look it up, we'd have learned that her odds of death were around one-

---

Australia 7; New Zealand 15; UK 12. And in a few sub-Saharan countries: Chad 1,100; Somalia 1,000; Sierra Leone 890

You can look up other countries using an interactive map of World Health Organization (2012) data here: http://www.who.int/gho/maternal_health/en/index.html.

* Thousands of babies have died too. Worldwide every year 3.3 million babies are stillborn and more than 4 million die less than a month after being born. Those babies die for many reasons, but it's clear that lack of access to medicine is a major factor because those deaths are heavily biased toward less developed countries (World Health Organization 2005).

eightieth of a percent. With those odds, we had the luxury to take control of the whole experience. We got to choose from among several options for the childbirth: Shelby could give birth at home or we could go to a hospital; we could have the baby delivered by a midwife or by a doctor; Shelby could use no painkillers at all or have the lower half of her body completely frozen by an epidural. No matter what our decisions, we knew that she and the baby would almost certainly both survive.

Those decisions, and others like them, were usually presented along a continuum between a "natural childbirth" and a "medicalized childbirth." Here's where I want to start challenging your assumptions about sex and nature. Take a moment to think about what defines a "natural" birth. Is it where the baby is born that matters? Is it whether or not drugs are used? Is it who delivers the baby that matters? What makes one childbirth more natural than another?

As Shelby and I got closer and closer to the due date, everyone kept telling us that we could make whatever decisions we wanted, but there was a definite pressure, especially from our friends, to choose the "natural" course wherever possible. People kept saying completely meaningless things like "Women have been having babies for centuries." (How does that help at all? They've been dying in childbirth too!) Worse, several people tried to calm us by saying things like "Nothing should go wrong just so long as Shelby relaxes, gets in touch with her body, and lets things take their natural course." Or "This is what her body was made for."

That is where my issue lies with the entire concept of a natural childbirth.

If you say, "Things go well when a mother is in touch with her natural self," you've implied that when things go wrong, it might be the woman's fault for failing to do so. It's sad enough that more than a quarter of a million women die from maternal causes each year,* but to blame even a fraction of them for their own deaths, because they somehow failed to be "natural" enough, is perverse.

In addition, that hijacking of the word *natural* puts a woman who does have access to modern technologies in a very difficult position. It erodes her ability to choose what she wants by making certain options at childbirth seem at odds with other decisions in her life. For example, if a woman chooses to eat organic foods, or if she enjoys spending time outside, she may feel pressure to make decisions about childbirth that are also labeled "natural." Of course, how a baby comes out of you has nothing to do with what you eat or how you exercise, but when things are presented that way, a person can be pushed into a decision that ultimately may not be the best one for them.

Take epidurals, for example. To say that a woman should be able to handle the pain of childbirth without an epidural because she's in touch with her natural body seems crazy to me. Giving birth is undoubtedly one of the toughest experiences a human being could ever have, so why should a woman be labeled "unnatural" if she uses modern medicine to help her through? I'm not saying all women should have epidurals. I'm just saying choosing an epidural shouldn't be seen as some kind of unwomanly act. Labeling an epidural-free childbirth as more "natural" than one with painkillers just isn't fair. Humans use drugs all the time, for all kinds of reasons, and humans have been using narcotics for centuries. Why is childbirth the only time drugs are suddenly so taboo? My hunch is that it's because women have epidurals and

* Around 287,000 women died in childbirth in 2010 (World Health Organization 2012).

men don't. If I have a headache, I take Tylenol, and nobody's ever called me unnatural for that. Why should a woman facing childbirth suddenly be forced to tough it out?

Then there's the whole decision about where to have a baby. Many people consider home births more natural than births in hospitals. That drives me nuts. Try to name one context in which the men who go to the hospital for a treatment are labeled unnatural. Creating an arbitrary and unjustified set of standards for women, on what might be the hardest day of their life, with no corresponding set of rules for men, is ludicrous. If it's so unnatural for a woman to go to a hospital, should we discourage women from working as doctors too?

Let me be clear: I'm not advocating for any particular choices—hospital or home, drug-free or epidural, midwife or doctor, C-section or vaginal, squatting on the floor or lying on your back. I don't even care if you want to eat the placenta afterward. Those are all perfectly valid options. I'm just saying let's take the false label of "natural" out of the equation so that women can be free to choose whatever experience they want.

The reason different people have different definitions of a "natural childbirth" is that it's an imaginary concept. Neither a hospital room nor a modern living room will replicate the places our ancestors delivered babies seven hundred years ago, and nothing from that era would match the experience our ancestors had a million years ago. While it might be helpful to think about the kinds of procedures women have used in the past, it's not fair to say a woman alive today can only be natural by trying to replicate what some women did long ago. And if they should do so, which women? In what culture? At what time?

As far as I'm concerned, it doesn't matter if a woman gives

birth on the international space station or if she gives birth squatting in the woods with her mother-in-law banging pots nearby to keep the bears away. If a baby comes out, it's a childbirth. Let's leave the word *natural* out of it.

Childbirth is tough for women, but there are animals out there that have it even worse than humans do. That's why I relish the moments when people say things about childbirth like "You should take inspiration from a pregnant cat or dog and the way they naturally know what to do, and just let nature take its course." People wouldn't say those kinds of things if they had pet hyenas, and it's fun to see the looks on their faces when I tell them why.

Hyenas are weird. They look like dogs, but they're not dogs at all. In fact, they're more closely related to cats than to dogs.* The female spotted hyena is especially weird, because she has a great big, hollow, tubular clitoris that really for all intents and purposes looks exactly like the penis of a male. It can even swell up and get erect just like a penis. The labia, which are the liplike structures on either side of a human's vaginal opening, have fused together in spotted hyenas and almost look like a scrotum. And as with a penis, the only opening of a female hyena's reproductive tract is at the very tip of that clitoris. They pee through it, have sex through it, and give birth through it. Even for biologists who work with spotted hyenas every day, it can be very difficult to tell a male from a female.

When a male sticks his penis into a female's clitoris during sex, her muscles pull the tip of the clitoris open, and then the

* The order Carnivora is split into two main groups: the "doglike" canimorphs and the "catlike" felimorphs. Canimorphs include dogs, bears, skunks, and sea lions. Felimorphs include cats, hyenas, fossas, and mongooses (Ignarsson et al. 2010).

whole thing folds outside in, kind of like the inverted sleeve of a shirt.* It's weird, but it works. Spotted hyenas mate and get pregnant without any major issues, but when the time comes for giving birth, that huge clitoris becomes a huge problem. The baby can't fit out the hole at the tip, and worse, the umbilical cord can't reach all the way from the placenta to the tip of the clitoris. The only way for the baby to come out alive is for the clitoris to tear open. That tearing happens the first time a mother gives birth. In captivity, 10 to 20 percent of females die while giving birth to their first litter. In the wild, survival rates appear to be much higher, but the clitoris tearing still happens. After that first birth, the torn clitoris heals, and the scar that's left behind seems to make subsequent births easier.[5]

Since they cause so much trouble, you're probably wondering why spotted hyenas have such ridiculous clitorises. It turns out it all comes down to spotted hyena society.[6] They live in a strict hierarchy, with an alpha hyena at the top, a beta hyena below that one, a gamma hyena next, and so on. What makes the spotted hyena hierarchy unusual compared to the group structures of other hierarchical mammals, like those of baboons, for example, is that the top rungs of spotted hyena society are held by females. In fact, every single adult female ranks more highly than every single adult male. When a female spotted hyena is born, she immediately outranks her father and can pick on him as a cub with impunity.

Females have a huge incentive to be aggressive. Females rank above males because they are larger and more aggressive than

* For a complete rundown on the mating habits of spotted hyenas, including an adorable picture of a male cuddling with a female after sex, see Szykman et al. (2007).

males are, and the higher a female's ranking is, the better her young's chances of survival will be. Furthermore, the alpha female has around 2.75 times as many daughters as the lowest-ranking female does. That's a huge difference, and it means aggression has a very direct benefit to the DNA controlling a hyena meat robot.

If you're a developing spotted hyena fetus, having an aggressive mom will be a huge help, but there's a drawback. One of the things that makes your mom aggressive is that her body is filled with hormones that make her act that way. Specifically, I'm talking about androgens, which are hormones usually found most abundantly in male mammals. (The word *androgen* literally means "man maker," so you may see where this is going.) As a result of being soaked in these hormones as a developing female fetus, the baby hyena ends up with male-like sex organs as a by-product. Apparently, females with tubular clitorises get enough of an advantage from the added aggression that it makes up for the costs they pay during childbirth.

I think the way spotted hyenas give birth should make it clear that there's nothing inherently safe or comfortable about that process. Evolution has favored aggressive hyenas, and the tubular clitorises that have evolved as a by-product are a life-threatening obstacle that females just have to deal with when they give birth. For humans, evolution has favored big brains, so laboring women are stuck squeezing giant-headed babies through the gap between the bones of their hips. There's no Mother Nature looking after hyenas or people. No one set up childbirth with any assurances of comfort or safety.

The issues that surround the idea of a "natural childbirth" echo many of those surrounding the whole debate about marriage equality. People on both sides of that debate invoke nature all the time, as though animal behavior were an instruction manual for how we should live our lives. On one side, people argue against gay marriage because they see examples in nature of a male and female coming together to reproduce. On the other side, people defend the rights of same-sex couples by pointing out that homosexual behaviors occur in many animals, including bats, penguins, and rabbits. Neither side is making a logical argument, though. There are plenty of exceptions to both of those rules, but more importantly, if you start handpicking behaviors from nature that you think should inform humans about how to act, you'll open a very ugly can of worms.

If humans used what happens in nature as a justification for human behavior, people could do anything they wanted to. Don't like the looks of that stranger walking through your neighborhood? Throw your feces at him. Are you worried the Joneses' kid is better-looking and more popular than your kid is? No problem. Kill the Joneses' kid, then eat the corpse for a boost in calories. Kick the people out of any house you want and move right on in. The world's your oyster.

Those examples might seem outrageous (okay, they are outrageous), but people really do use animal behavior as justification all the time. You might recall one infamous example of such reasoning that happened during the 2012 US elections, when Republican Senate candidate Todd Akin, of Missouri, got a lot of unwanted press after he spoke on a news show about abortion rights, rape, and women. He said, "If it's legitimate rape, the female body has ways to try to shut that whole thing down."[7] What a lot of people don't realize is that Akin was right about the female reproductive system—that is, assuming he meant "the female body of a duck."

The northern pintail is a duck that performs forced copulations, where males overpower females and force them to have sex. Males (drakes) accomplish those copulations thanks to their remarkably long penises. The length of the whole animal, from the tip of the bill to the end of the rump (without feathers) is just over twenty-three inches. The erect penis can be stretched out to a total length of around seven and a half inches. In human terms, that's a six-foot-tall man with a twenty-two-inch penis.[8]

When a pintail drake is swimming, walking, or flying around, you don't see how big that penis is because it sits outside in. During sex, though, a duck's penis inflates, spiraling counter-clockwise as it elongates into a corkscrew shape, and when it gets to full length it ejaculates. The best part is that the whole act takes about a third of a second.*

Blink your eyes once.

There. That took you about the same amount of time as it takes a duck to stick out its penis and ejaculate.

Now, if you look at a wild population of northern pintails, you'll find many more males than females, and that means a female northern pintail can be picky when selecting her mate. As a result, males have to compete aggressively with one another for a female's attention. Males will swim around a female, fighting one another for a position in front of her, then do highly ritualized dances—wagging their tails, shaking their heads, tucking the bill to the chest, and lifting their white breasts up out of the water to

* What's truly amazing about the duck's penis inflation system is that it's not driven by blood pressure, as it is for humans. Instead, that rapid expansion is driven by the lymphatic system—the same system that slowly drains excess water from your swollen tissues. The wonderful biomechanics paper that explains all this is Brennan et al. (2010). It includes some glorious high-speed videos of duck penises doing their thing, sometimes inside see-through glass vaginas. Those videos are also available on YouTube: http://www.youtube.com/watch?v=qwjEeI2SmiU.

show them off. Females generally prefer a male with a nice white breast and colorful iridescent feathers on the shoulders, and that dance is a great way for a male to show her what he's got. Once she chooses her male, the two of them pair up, head off together, and mate. She will have multiple clutches of eggs, so he will stick around to father them. He doesn't help raise the babies at all. He just hangs out until she wants to have sex again and in the meantime protects her from the advances of other males.

If a competing male manages to get past her chosen mate, that invader will attempt to force copulation upon her. That aggressive male's rapidly unfolding penis facilitates that process, by ensuring that he needs very little time to perform the act. His DNA's objective is to put copies of itself in her eggs. Her DNA's objective is to have the father of those eggs be the high-quality male she chose back when the males were doing their displays. That's why the male and female, even though they're mating, are in conflict.

The female avoids forced copulations whenever she can, but even when they do happen, she has a backup plan. Remember, his penis corkscrews counterclockwise as it enters her. The female reproductive tract has evolved in response to those forced copulations, so it contains eight clockwise spirals that lead to the eggs. That means his penis is spiraled the wrong way, and that makes it very difficult for a forceful drake to get his sperm where he wants them to go. On top of that, the vagina has three blind pouches where the tip of the penis can deposit sperm without those sperm ever getting near her eggs. This system isn't 100 percent effective, but far more of a female's eggs are fertilized by her chosen male than by the forceful drakes. Exactly how she guides the penis of her chosen mate to the correct place isn't known, but it's clear from anatomical investigations that her system does indeed have ways to "shut that whole thing down."

If you dissect a duck's vagina, you can see the corkscrew shape, and you can count the blind endings. In fact, if you look across duck species, you'll see that in species where forced copulations are more common, the females have more convoluted vaginas.[9] Female humans, however, don't have those blind endings. In fact, nothing about a human woman's reproductive tract suggests that she can control the outcome of a rape, except by using emergency contraception methods (i.e. taking a morning-after pill) or by choosing an abortion.

Rape is not acceptable among humans, even though forced copulation happens in animals all the time. And, of course, it *shouldn't* be. You can't use animal behavior to justify the horrible things people do to one another. Sometimes whole gangs of male ducks will force themselves upon a female until she drowns under the weight of them all[*] . . . and it's not just ducks. Gangs of two to three male bottlenose dolphins will surround a female for mating for days, biting and ramming her whenever she tries to escape.[†] A male squid will sneak up on a female in the darkness of the ocean and fire his sperm packets into her as suddenly and rapidly as he strikes prey when he feeds.[‡]

---

* Female waterfowl are especially likely to be killed by marauding males in parks, compared with in the wild, because the sex ratio in parks is often even more biased toward males than usual (McKinney and Evarts 1997).

† Yes, even Flipper had a dark side (Connor et al. 1992).

‡ It's dark where deep-sea squid live, and it's hard to tell males from females, so when a male encounters another member of his species, he injects a sperm packet into its skin as quickly as possible. If it's a female, he'll probably fertilize some of her eggs. If not, well, no big deal, really. Researchers cleverly named this a "shot in the dark" strategy (Hoving et al. 2011).

Forced copulation is everywhere, and the reason it's so common is that even though a male and a female might be mates, they can still fundamentally be in competition *against* one another. I've already shown you that animals will injure their own bodies to reproduce, so it should come as no surprise that they're willing to hurt their mates too. The basic rule of the selfish, Scrooge-like game still applies: the only question that matters to an animal is "How can *my own* DNA get passed on?" When that's an animal's motivation (as opposed to romance or love), things like infidelity, physical violence, and forced copulation are bound to happen.

Sometimes even what looks like consensual sex can be a case of forced copulation. For decades it was assumed that garter snakes had consensual sex. When garter snakes mate, a male lies on top of a female and puts his penis—which is forked, by the way—into her cloaca. (Her cloaca, just for the record, is the single hole in her body that urine, feces, and eggs come out of, and that a male's penis must enter during sex.) Before he does this, though, he lies against her and makes waves with his body that ripple upward, from her tail toward her head. People had previously interpreted this as his way of turning her on so that she would open her cloaca to him.[10]

However, about a decade ago, researchers realized that if you make waves like that against a female snake's body, it pumps all the air out of her lungs and stops her from taking a breath. As she starts to run out of oxygen, she gets stressed, just as she would if a predator were attacking her. So instinctively, she craps herself. That reaction might discourage some predators from eating her, but unfortunately, it does not repel the male. Instead, it opens her cloaca and he takes advantage by inserting his forked penis.

As horrible as the sex lives of garter snakes and northern pintails may be, they have nothing on bedbugs. Bedbugs are those parasitic insects that bite you when you're sleeping. They used to be something you only had to worry about in dingy hotels (à la Henry Miller's *Tropic of Cancer*), but in the past decade or so their populations have skyrocketed around the world, and today you can even pick them up in five-star hotels. The reason for their resurgence is unknown, but it's quite alarming. Bedbugs feed on human blood, but however disturbing their feeding habits may be, those don't hold a candle to how gross their mating behavior is. Bedbugs are *nasty*.[11]

Bedbugs are tiny, and they hide most of the time in walls, electrical sockets, piles of clothes, or pretty much anywhere in a bedroom. About once a week, a female bedbug comes out of her hiding place, crawls into bed with a human, bites the person, feeds for twenty to thirty minutes, and then crawls back to her hiding place. As she heads back across the room to hide, she's well fed and therefore at her most fertile. So, on her way back she is approached by several males.

This is when things get ugly.

Bedbugs have a bizarre form of mating called traumatic insemination, in which the male pierces the body wall of the female with his penis and injects his sperm into her body cavity. She has a hole, through which she lays eggs, but he doesn't put his penis in there. Oh no—instead, he just stabs her in the middle of her body and makes a new hole. This obviously damages the female, reducing her lifespan about 30 percent below that of a virgin female. Too much mating can even kill her.

Now, a female needs to mate in order to pass on her genes, and traumatic insemination is the only ticket to pregnancy, so mating is helpful to her, even if she does get stabbed in the process. Her problem, though, is that it doesn't just happen once. On her journey from the human's bed to her hiding place in the wall, she will typically receive traumatic inseminations from five or so different males. One mating would give her more than enough sperm to fertilize all her eggs, but five traumatic inseminations give her twenty to twenty-five times more than she needs, and each one of those matings further reduces her lifespan. For a female, it's a bad situation.

Males don't care about that, though. To a male, all that matters is that *his* sperm are the ones that fertilize her eggs, and as a general rule, the last male to mate with the female gets to be the father of about 68 percent of the offspring. So even if she has mated with four other males today, and even if another mating will push her ever closer to the brink of death, there's a huge incentive for a new male to be the fifth.

For her part, the female can't really do very much about the traumatic inseminations themselves. Instead, females have evolved an organ called a spermalege to help them deal with the aftermath. Mostly it brings immune cells to the stab wound to make sure the fungi and bacteria all over a male's penis don't infect her body, but the spermalege also has a role in moving the sperm to the part of her body where the eggs are. Traumatic insemination by multiple males clearly isn't optimal for her, but her body does what it can to deal with the brutal reality of life as a bedbug.

Forced copulation and traumatic insemination are instances where males hold the upper hand in the conflict between the sexes, but there are plenty of other examples where females have made things just as rough for males. Where females have a little more control over matings, they can choose some feature in males to judge them on, and then only mate with the males who meet their standards. Unfortunately for the males, females tend to choose a male feature that is most likely to get him killed.

For example, a female túngara frog will not mate with a male unless he woos her in the night with a loud, musical call.[12] As a result, if you go out into the rainforests of Panama at night, you can hear males of the species calling as loudly and deeply as they can, each trying to attract a female. The problem is, other animals can hear that call too. One such eavesdropper is one of my favorite animals from that part of the world, a bat called *Trachops* or, more commonly, the frog-eating bat. That bat finds its food—yes, frogs—by listening for their calls.*

Calling out into the night like that is basically a male túngara frog's way of shouting "Eat me!" at the top of his lungs, but female túngara frogs have left him no other choice. Her logic is this: If a male can still survive despite the constant risk of bat attacks, he's probably got great DNA. Any males that don't take that risk aren't worth mating with, and as for those males who get eaten by the bats? Well, they didn't have very good DNA, now, did they?

This system of evaluation by the females can only work if they choose a task for males that really is dangerous. If they picked something easy, every male would do it equally well, so the females would end up mating with crappy males just as often as with good males. Females who chose an easy contest for males would therefore get no benefit from holding the contest in the

* One thing I love about these bats is that they can tell the difference between the calls of poisonous and nonpoisonous frogs, so they can selectively attack nonpoisonous frogs and ignore the calls of poisonous ones (Page and Ryan 2005). Unfortunately for the male túngara frog, he is not poisonous.

first place. As a result, across species, the kinds of hoops males have to jump through to have sex are insane. The most studied examples come from birds.

Bright-colored birds doing intricate dances while singing loudly are beautiful to us, but when you think about it, those males have to use a ton of energy to keep themselves beautiful and perform those displays properly.* Besides, those displays have them basically doing everything they can do to get noticed by potential predators. Also, the bright, beautiful red feathers many males display are attractive to females because those yellow, orange, and red pigments are very metabolically expensive to produce, thus making it harder for the male who produced them to survive.†

Female spiders who eat their mates are another obvious example of females getting the upper hand on males, but male spiders often pay a high price before they get anywhere near a potential mate. For example, female golden silk spiders sit in their webs, merrily dining on insects, waiting for males to find them. While a female has only a 0.3 percent chance of being eaten by a predator on any given day, a male roaming the rainforests of Panama in search of her has about an 8 percent chance of dying each day. That's more than twenty-six times higher. Once he finds her, they mate, and then she tries to eat him anyway.‡

---

* The most spectacular male birds you could ever possibly see display are the birds of paradise (Scholes 2008). I'm talking about the actual birds, not the flowers with the same name. If you're not familiar with birds of paradise, spend five minutes on YouTube to see some of the videos Ed Scholes and Tim Laman have taken. Your eyes will explode out of your head in disbelief. Here's a start: http://www.youtube.com/watch?v=-YTR21os8gTA.

† Those red, yellow, and orange pigments are called carotenoids (Hill 2000).

‡ This difference in life history between males and females is part of the reason females

Some females have done away with males altogether, becoming all-female species that use cloning to reproduce instead of sex. This has happened in more than eighty different species, including New Mexico whiptail lizards, several fishes, and a few salamanders. This is a fascinating phenomenon to start with, but it's particularly interesting in some of the female-only salamanders, in which the females have absolutely no genetic use for males but still need sperm to stimulate the development of their eggs. In other words, they're looking for a male to have sex with, but the male will get no genetic benefit whatsoever from the mating.[13]

This is such a rotten deal for the males of the species that . . . well . . . there aren't any males. That's what happens when you can't pass on your genes. But that hasn't slowed the females down. To get the sperm they need, females just have sex with males of other salamander species. That stimulates the development of eggs, so the females can clone themselves successfully. The babies aren't hybrids of the two species, they're just female clones of their mom, and the salamander she had sex with gets no benefit from the mating at all.*

Some animals out there possess both male and female parts, but that doesn't make sex any easier on them. One striking example is the flatworm *Pseudobiceros*. Its name means "false two horns," and you'll see why in a moment. There are several different species of

---

can afford to be so big and noticeable, while males need to be small. Their size helps them avoid getting spotted and eaten while they look for females (Vollrath 1998).

* Well, he might get the benefit of having an enjoyable time, but he doesn't make any babies this way. From his DNA's perspective it's a waste of energy and a waste of sperm. To get any real advantage from mating, he needs to mate with a member of his own species.

Pseudobiceros, all of which occur in the Great Barrier Reef of Australia. They're only a few centimeters long, but they're quite beautiful, like tiny undulating oval-shaped magic carpets, some with bright frills of color along their sides. All *Pseudobiceros* have two short bumps on their bellies. Those bumps are what give these flatworms their name. As you know from their Latin name, those aren't really horns. Yup. They're penises.

Each flatworm is a hermaphrodite—it has the penises but also the complete reproductive system of a female. That means that when two *Pseudobiceros* flatworms rendezvous to mate, they have to decide who will be the father and who will be the mother. The mother will be pregnant, which requires a lot of energy. The father only has to ejaculate. Since they're both selfish, neither of them wants to get pregnant, so how do they settle it? Simple.

It's called penis fencing.[14]

The two animals wrestle, parrying with their penises until one stabs the other with one of its two penises. The winner injects sperm into the other's body. The loser gets pregnant. Problem solved.

If you look across all these lusty examples, it becomes clear that sex has a considerable cost for males and females. With males forcing females to copulate, females forcing males to put their lives on the line, and hermaphrodites fighting over who gets to be the daddy, it's a wonder these animals bother at all. So why haven't animals abandoned sex altogether?

There was a time on Earth before sex, when everything reproduced by cloning. Back then, there were no males or females; there weren't even really parents and children. There were just organisms. Everyone had a single cell and lived in the water. Once in a while a cell would split in half, and the two cells that resulted would float apart. You couldn't really call one of them the parent and one of them the child. They were just clones. For a selfish DNA strand, cloning worked perfectly well.

Then sex evolved, about 1 billion years ago, in a small group of single-celled organisms.[15] It was such a successful strategy that the descendants of those creatures thrived. Today, every single plant, fungus, and animal on Earth is a descendant of those first sexual creatures, and sex remains an integral part of their continued success.

Sex is everywhere. There are a few animals that use cloning to supplement their sex lives, like jellyfish, starfish, and some worms, but only a tiny minority of animals have left sex behind altogether, to use cloning exclusively. Since asexual animals are so rare, one of the big questions biologists love to ask is, Why? What's so great about sex? After all, sex gives you offspring with only 50 percent of your own DNA, instead of the 100 percent you get from cloning. So why is sex so popular?

So far as we can tell, the answer is twofold. First, sex helps an animal's offspring deal with an unpredictable world better than clones would. If you have multiple kids from sex, they'll be different from one another, the same way you're different from your sisters and brothers. Let's say, down the road, there's a flood. Well, that might suddenly give an advantage to animals that are a little stronger, better at climbing, or better able to hold their breath. But if instead of a flood, there's some new predator to deal with, the advantage might instead go to animals that run faster or are better camouflaged or maybe are just stinkier. There's no way to know what the future holds, so having a diverse set of kids is the best way for a parent to end up with at least one kid who has the right stuff. It's like spreading your bets across the roulette table. With cloning, all your kids would be identical. Sex shuffles the genetic deck in a way that cloning does not, and that's useful in a changing and unpredictable world.

The second great advantage of sex over cloning is that it gives the parent an opportunity to improve its DNA's future odds of survival by mixing it with the DNA of someone else. Until sex evolved, cloning animals had no other option than to simply pass on whatever DNA they had been born with. With sex, an animal can look around for someone that looks like they're doing well in the world, and then, by having sex with them (as opposed to some other animal), bring those favorable genes into the family. That's why sex was a total game changer for life on Earth. Cloning animals can ignore one another, but with sex, everyone has to constantly check one another out. Your DNA doesn't just want you to mate, it wants you to mate with the best possible partner (or partners) you can find. Whether or not your own babies will be able to compete will largely depend on the quality of your mate's DNA. It's not an exaggeration to say that your sex life is just as important to your DNA as your survival.

Ultimately, this urge to find great DNA forms the basis of the whole dating game. For humans, romantic love is an integral part of that, but romance isn't really "natural" in that it's not part of most animals' sex lives. Raising a human baby to the point of self-sufficiency takes more than a decade, and having both parents work as a team to care for that child is a great strategy to ensure its success. As human meat robots, we have urges written into our DNA to help motivate us to work as partners with the people we have sex with. I would argue that the long childhood stage of humans is the only reason we have romantic love at all. Humans are still fundamentally selfish, for all the same reasons that other animals are selfish. But because human kids take so long to raise, a person can still be a good partner without giving up on the selfishness that goes along with being an animal.

I think it's fun to look at my relationship with Sam's mother in this context. When I first met Shelby, I had an urge to talk to her. When I talked to her, I wanted to spend more time with her, and then spending time with her made me want to do the things that my DNA had in mind in the first place. Meat robot, meat robot, meat robot.

Shelby impresses the hell out of me: she's the kind of person who does things most people only ever forward emails about. She helped with the environmental cleanup after Hurricane Katrina in Biloxi, Mississippi. During grad school she performed with a professional dance company in Providence, Rhode Island, and before that she produced a dance show for the Minneapolis Fringe festival. She was even in one of YouTube's most viral videos ever—the one from 2009 where the whole wedding party dances down the aisle, called "JK Wedding Entrance Dance."*

Shelby's also a scientist like me, and while we were dating, she did the fieldwork for her PhD in the Amazon rainforest of Brazil.† She had no phone or Internet connection at her field site, so she'd make her way to the nearest town every other weekend, get a hotel room, and use Skype to fill me in. I was so jealous: she was there to collect soil and water samples but ended up seeing

* The link to that video is here: http://www.youtube.com/watch?v=4-94JhLEiN0. Last time I checked, it had more than 80 million hits.

† Shelby's PhD was mostly about what happens to the quality of water and soil when you convert native Amazonian rainforest to a giant soy field (Riskin et al. 2013).

jaguars, pumas, anacondas, beetles, ants, giant anteaters, sloths, tapirs, rattlesnakes, and even some bats. That said, though, she also experienced nature's rougher side while she was there, and I'm not quite so jealous of those experiences. She was attacked by swarms of bees, which got tangled in her hair as she ran away, and then kept stinging her while she tried to get them out. That happened three times. On another morning, she couldn't quite seem to get her boot on all the way because something at the bottom was pressing against her toes. She removed the boot, slammed it against a wall, and a giant tarantula fell out of it. She once drank from the wrong tap at her field station and got stomach parasites called *Giardia* that made her so sick that, even though she nuked them with antibiotics, her digestive system didn't get back to normal until about a year later.*

Shelby and I had lived together for only a few years before we decided we wanted to have a baby together—that is to say, before our DNA molecules coerced us to mate so that they could replicate. My body cut out twenty-three double-helix strings of DNA and placed them into a sperm cell. Shelby's body, through an identical process, had already placed twenty-three of her own DNA strings into an egg. Then we had a special kind of sleep (wink, wink), and in the instant those two cells met, they both vanished. In their place was left a unique new being with (23 + 23 =) forty-six strings.

* The *Giardia* parasite is one you can also get in North America. You might have heard of it by its nickname, "beaver fever." It can cause intense (and I mean *intense*) abdominal pain, diarrhea, blood in the feces, and sometimes even blood in the urine. Humans get *Giardia* by ingesting water tainted by the feces of animals (beavers, humans, whatever) that are already infected by the parasite. In other words it's completely preventable so long as people have access to clean water. Despite that preventability, hundreds of millions of people worldwide have that parasite right now.

Suddenly, Sam was.

At first, Sam didn't have a smile or eyes or even limbs. He was just a single cell, separated from the outside world by nothing more than a thin, oily, spherical layer—almost like a tiny soap bubble. You'd be able to see that bubble with your naked eye, but only just barely, and the DNA strands inside would be invisibly small. The sequence written into those strands, though, instantly made Sam absolutely unique: the sequence of DNA written across his forty-six strings has never existed before in the entire history of life. And it wasn't just a jumble of letters either. It had meaning. That DNA sequence encoded the building instructions for the meat robot that now exists as Sam.

Shelby and I didn't really create a life: our own cells were already alive when they merged. Sam is the next link in an unbroken chain of constant life that stretches backward through time—through his parents, through his grandparents, and beyond. Because each of Sam's ancestors had a different sequence of DNA, the bodies built by their DNA molecules varied too. Sam's body is similar to mine, and mine is similar to my father's, but those small changes begin to accumulate as the story stretches back. Sam's story goes back through me, through Ice Age hunter-gatherers, through apes walking on all fours, through squirrel-sized tree-climbing primates, through prehistoric reptiles, through alligator-sized amphibians, through strange, lobe-finned fishes in the ancient ocean, through wormlike protofishes in the ocean before that, and so on. In fact, by the time you go all the way back to the beginning of that chain, to the beginning of sex itself, there's not much more to look at than a simple bag of DNA molecules floating in the primordial ocean—a single cell that looked an awful lot like Sam did on the first day of his life. All that's really changed over those billion years are the meat robots that the DNA molecules have become able to build. The DNA itself has stayed pretty much the same that whole time.

But now he's a child. When I hold him, Sam feels like a miracle. I love Sam—*love* him—and that's helped me understand how DNA molecules coerce their meat robots to do their bidding. I'm experiencing the most powerful emotions of my life. If I were a bowerbird, I'd have decorated my nest with blue objects. If I were a bighorn sheep, I'd have smashed my head into anyone I could. But I'm a human, and I instinctively care more about Sammy than anything else in the whole world.

As for Shelby and me, we feel like a team. We look after one another, we support one another, and we're happy together. Despite the horrible things animals do to one another in the name of sex, we've found a way to leave that conflict behind. You can call that natural, but it's not really. It's human.

Overall, sex has made me a pretty happy guy. Even if the testosterone that comes with that joy has cut my life shorter by a decade or two, I would say it's been worth it.

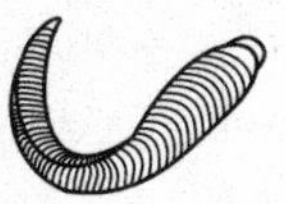

3

SLOTH

Just Another Day in Parasite Paradise

Sloth, or laziness, seems at first like something that only affects humans. After all, we invented TVs, recliners, and video games. We work at desks, we drive cars, and we take the elevator when it's only one floor. Obesity is on the rise in almost every country now. By 2008, 1.5 billion adults were overweight, 170 million children were overweight or obese, and those numbers continue to climb.[1]

Just as you'd expect with all these big people walking around, experts have been coming out of the woodwork with advice on how to lose weight and stay fit. Many of those antidotes have to do with living "naturally." Our ancestors weren't fat slobs (so far as we know), so we're told that our lost connection with nature is what's made us so lazy. Nature's a perfect model of hard work, isn't it? Survival of the fittest, right?

Nice try.

Sure, I'll concede that a beaver, for example, epitomizes industriousness and hard work as she builds her dam. But beavers aren't the only animals out there. The belly of that beaver is home to a whole ecosystem of lazy creatures, stealing mouthfuls of her food for themselves, some even feeding on the beaver's own flesh. Those freeloading parasites thrive inside that beaver's body, taking full advantage of all her hard work. And they give absolutely nothing in return.

You cannot possibly get any lazier than that.

If you ask, most people will tell you that they love nature, but many of those same people don't even want to look at a picture of a parasite. Fact is, though, parasites are as natural as the birds and the bees. In fact, birds and bees are covered with parasites of their own, as is pretty much any other animal you can name. In fact, although it may sound counterintuitive, biologists have even argued that parasites are the sign of a *healthy* ecosystem.[2]

Like most people, I didn't love parasites at first. I was curious about them, but I went into my undergrad program to study bats. The only reason I learned anything at all about parasites was to fulfill my degree requirements. Parasites were confusing to me. I can still remember the agony of trying to memorize their Latin names and life cycles and trying to tell them apart in their specimen jars. They honestly all just looked like overcooked spaghetti. I learned what I needed to pass the courses, but the parasites never really engaged me.

I would have laughed if someone had told me then that in just a few years I'd appear on an internationally broadcast Animal Planet TV show about parasites called *Monsters Inside Me*, or that

I'd become a frequent guest on the *Late Late Show with Craig Ferguson*, talking with him about the very same worms whose names I previously couldn't even remember. In a few short years, parasites would become a big part of my life, but before all that, I had to somehow realize how incredible they were.

Fortunately, there's a bat for that.

Within a month of starting my master's degree, I was sent by my advisor down to Costa Rica to figure out whether a project idea would have any chance of working.* It was my first time in the tropics, and there were so many bats to see. On my second day there, I went with a more senior graduate student named Maarten Vonhof to look for bats in a small cave. That was a very good day—my first encounter with vampires.

Vampire bats drink the blood of other animals, so they are parasites. By definition, a parasite is an animal with a relationship to some other animal, called a host. In that relationship, the parasite gets an overall benefit, and the host pays an overall price. The parasite might get nutrition, protection, or transportation. The host might have some food stolen, or it might get injured, or it might even die. Whatever the exchange, the parasite comes out ahead and the host loses. It doesn't matter if the parasite is

* The project would ultimately be designed to figure out how bats with these little suction cups on their wrists and ankles use them to hold on to the smooth surfaces of leaves. A couple of years later, I published that paper (Riskin and Fenton 2001) and then followed up with a second paper a decade later, about a similar bat that lives in Madagascar and does the same thing (Riskin and Racey 2010). These two different kinds of bats are a remarkable example of convergent evolution, having evolved their adhesive organs separately. The Costa Rican bat sticks to leaves via suction, but the one in Madagascar uses wet adhesion, like a wet piece of paper sticking to glass. It took me more than twelve years to uncover all that, and it all began with that first trip to Costa Rica.

a worm, a fish, or a bat. So long as it's mooching off a host, it's a parasite.*

Vampire bats eat blood and nothing else. Although that might seem grotesque, it's really kind of brilliant. If you think about it, blood is the perfect food. When you eat, your digestive tract pulls nutrients out of your burrito and puts them straight into the bloodstream so that as the blood flows through your body, all your cells can get those nutrients. When a vampire bat drinks a cow's blood, it taps into a cocktail of molecules that includes everything it needs. That said, getting blood out of a cow that weighs 14,000 times what you do isn't easy.

Of the more than 1,200 bat species that live in the world, only three drink blood. They don't turn into well-dressed European men with fangs, but they are called vampire bats nonetheless. All three species live in Central and South America, with two of them feeding on the blood of sleeping birds, by sneaking up on them in trees and biting their toes, and the third, called the common vampire bat, feeding on the blood of mammals. While the common vampire bat usually feeds on cow blood, it can feed on a wide range of animals, and it has been known to feed on sleeping humans.†

* It's actually possible to drink blood without being a parasite, just so long as you find a way to do it without hurting your host. For example, the spider *Evarcha culicivora* of Kenya feeds on human blood, but it never harms humans. It gets all the human blood it needs by killing and eating mosquitoes. Mosquitoes that haven't fed on humans don't have enough nutrition for the spider—it really needs the blood—but since it takes that blood from humans without ever hurting a human, it's not a parasite of humans. If anything, since it eats mosquitoes, the spider that drinks our blood is an ally (Cross and Jackson 2012).

† Common vampire bats almost always feed on cattle, and that presents a wonderful mystery: since vampire bats only live in Central and South America, and since there were no cows in that part of the world before 1492, no one knows what vampire bats fed on before livestock was brought from Europe. Biologists assume that vampire bats originally fed on all kinds of different rainforest mammals, and that they switched over to cattle once the cattle were introduced. Imagine eking out an existence drink-

The common vampire bat is about the size of a mouse.* To feed, it sneaks up on a sleeping cow by crawling up to it on the ground. When it gets close enough, it uses heat sensors on its nose to find blood vessels close to the cow's skin. It's not looking for a big jugular vein or anything, just some capillaries: think of your cheeks, the top of your head, or your fingers and toes. Places that get rosy from blood just below the surface, and that feel warm to the touch. On a cow, capillary beds like that can be found around a hoof, on the neck, around an ear, or right on the sex organs. That's where vampires bite them.

Unlike Dracula, the real vampires don't clamp on with fangs. Instead, the bat neatly shaves fur with its teeth from the area it wants to bite, then makes a shallow slash in the skin with its upper middle two teeth. The resulting divot is only about a quarter inch deep and a quarter inch wide, but it bleeds, just like a man's cheek does when he cuts the capillaries in his face while shaving. Thanks to compounds in a vampire bat's saliva, the divot in the cow's skin will continue slowly dripping blood for as long as the feeding takes.[3] The bat will feed for twenty to forty minutes, urinating all the while to rid itself of excess water, until it's consumed about a tablespoon of blood, increasing its weight by around 50 percent. Then the bat jumps into flight and goes back to its roost to hang out.

It's an important point that vampire bats approach their hosts by walking up to them, because walking is super weird for

---

ing blood from well-hidden, small rainforest mammals that can bite you back, and then suddenly having giant, defenseless cattle brought in. It would have been like getting a McDonald's drive-through in your neighborhood. Today, it's almost impossible to find common vampire bats that don't get all their food from cattle, so we may never know what the original food of vampire bats was.

* A non–Gough Island mouse, that is.

a bat. Vampires only walk because, unlike all the other bats, they make their living on blood. Most bats never land on the ground at all, and those that fall accidentally typically jump off the ground as soon as they can. Vampire bats, on the other hand, are perfectly comfortable walking on the ground, and they can launch themselves off the forest floor with amazing dexterity. In fact, there's a very famous biomechanical study on vampire jumping; it shows that a vampire bat can launch itself in less time than it takes for a human to blink (which, as you already know, is about as fast as a duck unfolds its penis) and can jump upward three feet or more. For an animal the size of a mouse, that's insane.[4]

Hiking toward a cave in Costa Rica that I knew might have vampire bats in it made me feel like a twelve-year-old going backstage at a Justin Bieber concert. Maarten, who had been in Costa Rica for several months already, said that the vampires had been in the cave each time he'd checked, but when he saw how excited I was getting, he was quick to mention that he hadn't been to the cave in several weeks, so there was no guarantee. First, we canoed across a small river, tied the boat to a branch, then set off along a path. As we walked, the forest rose up high above me, and every bump on every branch had a very real possibility of revealing itself to be a monkey or sloth. It was muggy and wet and smelled like mud with a hint of spice. There was a constant buzz of cicadas, and I could see insects all over the forest floor, but there were surprisingly few mosquitoes. Colorful birds zipped across the path ahead of us. I'd always wondered what tropical rainforests were like, and now I was in one.

We stopped about thirty feet from the cave, and Maarten explained the layout so that I'd know what to look for before scaring all the bats away. On the wall near the mouth of the cave I'd find some sac-winged bats. Those are insect-eating bats best known for the way males throw urine, saliva, and semen on females they want to mate with (more about them in the chapter on envy). Beyond those bats, I would need to crouch to enter the main chamber, which was only about four feet high. On the ceiling, I'd see some short-tailed fruit bats—bats that can tell what kinds of nearby fruit species are ripe by smelling the breath of well-fed bats coming into their cave. Past that main chamber, Maarten told me, the cave narrowed to a hole in the back wall at floor level, some eight feet or so from the entrance. That hole had an open shaft above it, almost like a chimney above a fireplace. Maarten told me that if I slid on my back, pushed my head into that hole, and looked up, I might see a few vampire bats three or four feet directly above my face.

I walked to the cave as silently as I could, found the sac-winged bats, and took some photos. Then I crouched to look in the cave. There was movement, but I was still too tall to really see what was going on, so I took off my backpack, turned around, then lay down on my back and slid into the cave. The cave floor was wet and smelled *terrible*. As I slid through the bat droppings, I felt the sludge slide through my hair, down my neck, and into my shirt. Halfway into the cave, I turned on my headlamp and immediately saw the short-tailed fruit bats, huddled together, looking straight down at me. One of them flew away when my light came on, but then it came right back just a few seconds later and landed right where it had been, next to another two. As they hung, they twisted by their toes, pivoting around, with their faces rotating above me. They were really looking at *me*, and although I couldn't hear it, I knew they were observing me with their ultrasonic echolocation calls as well. I'd never experienced anything like this. It was the kind of moment I'd dreamed of having ever since I read my first book on bats in high school. I would have preferred not to have insect-laden guano down my neck, but in spite of that, this was the coolest thing I'd ever done.*

A vampire bat needs to feed pretty much every night or it will starve to death, so if one comes back to the roost without having successfully found a meal, it will beg the other vampire bats to barf up blood for it. The hungry bat does this by going from bat to bat, licking their mouths until one of them pukes a little to help them out. What makes that French-kissing blood-vomit exchange so incredible (besides the fact that it's a French-kissing blood-vomit exchange) is that bats will even help unrelated individuals this way. In other words, while it wouldn't be surprising to see a mother feed her baby, or even a brother help a sister, you don't ever expect animals to help nonfamily members. That would seem to violate the Scrooge-like rules of selfishness that govern all animals. But the food sharing by vampire bats works because the vampire bats are smart enough to remember who has helped in the past and who has not. If a bat keeps on begging but never shares, the group can just stop puking for that one bat. By puking today, a bat ensures that it will get help when it's down on its luck next time. Other than humans, vampire bats are among the few animals on Earth that will give away food they have already secured just to help out an unrelated member of their own species.[5]

---

* The book was *Just Bats* by Brock Fenton (1983). After I read it, I contacted Brock and told him I was interested in learning more about bats. He was very supportive. Years later, he invited me to do my master's degree with him. He was the one who sent me on this very trip to Costa Rica.

I turned off my headlamp and slid onward through the dark, until my head went into that hole at the back of the cave. As I slid from a floor covered in pooped-out fruit to a floor pasted with pooped-out blood, the smell got worse. I came to a stop; then I heard the loud screeching noises coming from above me—sounds I'd never heard, but that I could only assume were the voices of vampire bats! I brought my right arm up toward my head to turn my headlamp on, but my knuckles smacked into rock. That's when I realized that my head was plugging the only exit to the vampire bats' chamber. It was too dark to really see what was going on.

I tried a different path for my arm, up my chest and against my face, and was finally able to flick on the headlamp.

There were three of them!

I knew those faces instantly: common vampire bats, *Desmodus rotundus*. In the light they got louder and started moving around on their perch above me. They bared their pointy triangular teeth at me and screeched like miniature dragons. I couldn't get my camera into the hole with my head, so I had to just lie there and soak it all in. I fully admit that it was uncomfortable—even scary—but I think that was part of what made the whole experience so life-changing. I wasn't in control. My face was exposed to them. The best way I can think of to describe the vulnerability is to say that it was like swimming in the ocean for the first time. You're scared, but the thrill and the beauty are enough that you're *willing* to be scared.

These three vampire bats were exotic, but what made them even more beautiful to me was the context—all those scientific facts I'd acquired made them familiar, like celebrities. I knew so many things about vampire bats that set them apart from all the other bats. Feeding only on blood, with no variety in diet at all, has caused vampire bats to lose their ability to discern different flavors.* Because they drink up to half their own body weight in one meal, the stomach of a vampire bat has a side pouch that can fill up with blood quickly and then trickle that meal through the digestive system slowly over the ensuing hours.[6] That same out-pocket is what makes it possible for bats to easily puke blood up for one another when they share food. By extension, the food-sharing behavior itself is a by-product of the fact that vampire bats have become parasites.

What started to dawn on me that day was that the vampires are strange and charismatic *because* they feed on blood. In other words, being *parasites* is what made vampire bats so wonderful. Lying in that cave was one of the critical moments that set me on the road to becoming a bat biologist, but it was also the first time I really appreciated how incredible parasites could be. That sparked a curiosity in me about whether there might be any other lazy parasites out there worth a second look.

Were. There. Ever.

Parasites live everywhere there is life. The squirrel in your backyard is filled with microscopic creatures; the bird at your feeder is covered in teeny parasitic arachnids. Panda bears have parasites, as do the penguins of Antarctica.[7] In fact, to my knowledge, biologists have never found an animal species *without* parasites. Those lazy parasitic lifestyles are hugely successful. If a full census of life on Earth could ever be completed, biologists have

* The fact that vampire bats can't taste anything was uncovered by a guy with whom I did my master's, John Ratcliffe. The experiment is pretty simple: you feed an animal something with a strong flavor, and then you make it feel sick. Afterward, it won't want to eat anything that tastes like that. You know how your uncle Bert can't stand the smell of tequila because of a bad experience he had in Mexico that one time? Same thing. To date, vampire bats are the only animals ever found that don't learn to avoid a particular taste after repeated experiences of sickness (Ratcliffe et al. 2003).

argued, the total number of parasite species would likely be even bigger than the number of nonparasite species.[8]

Humans can get parasites too. You've experienced a mosquito bite, I'm sure. Remember the itch? That happened because when that mosquito bit you, she began by spitting into your skin.* She hawked that loogie into your blood before she started drinking because her spit prevents your blood from clotting. Your immune system went into action to clean up the spit (too slowly to stop the mosquito, but fast enough to make you itchy for a few days), and the inflammation that resulted was itchy.[9]

In most cases, the mosquito bite has no aftermath, but sometimes a mosquito's salivary glands are home to parasites called *Plasmodium*. Those are tiny wormlike creatures, even smaller than a single red blood cell. When a *Plasmodium*-infected mosquito bites you, she squirts those parasites into your blood. They float through the bloodstream to the liver, embed there, and breed. After some time, they leave the liver, float back into the blood, and then somehow find their way to the insides of red blood cells—where they reproduce, then burst out, destroying the red blood cells. That process gives the human host horrible fevers, liver damage, and sometimes even inflammation around the brain and spinal cord.

The disease caused by *Plasmodium* is called malaria, and no other parasite (or predator, for that matter) causes as much human pain and suffering as those little wormlike bastards in the salivary glands of mosquitoes. Using those mosquitoes as their delivery system, *Plasmodium* parasites infect hundreds of millions of people each year, killing thousands of them, mostly children.[10]

---

* I say "she" because only female mosquitoes bite humans. Mosquitoes of both sexes eat fruit juices and nectar, but females also feed on blood to get iron and a few extra proteins for laying eggs.

If having worms enter your red blood cells isn't bad enough for you, Mother Nature has a smorgasbord of other human parasites to choose from, each more disgusting than the last. For example, I'm personally quite grossed out by the roundworms that cause that grotesque swelling of the lower body called elephantiasis.[11] You get those worms from a mosquito bite, and they quickly set up shop in your lymphatic ducts. (Lymphatic ducts are tubes that drain excess fluid from swollen tissues and put that fluid back into the bloodstream.) Once the worms find those lymphatic ducts, they grow to somewhere between one and four inches in length and hide, undetected by the immune system for up to thirty years.

Thirty years!

Do you realize how long that is? Hiding a four-inch worm inside your body without the immune system going nuts should be impossible, even for a day. If humans could mimic the cloaking ability of those worms, it would be an absolute game changer for organ transplantation, where rejection by the immune system is the major barrier to survival. I'm not saying we should love these worms (I sure don't), but you've got to respect what they can do.

Eventually, the worms will die of old age, and when that finally happens, the cloaking system shuts down and the human immune system, suddenly realizing there are giant worms in the body, goes nuts. Swelling is intense and immediate, but the lymphatic ducts that should drain that excess fluid are blocked by the corpses of the worms. As a result, the limbs swell up to unimaginable dimensions. Voilà. Elephantiasis.

And just in case elephantiasis and malaria aren't enough to convince you of Mother Nature's dark side . . . You can get roundworms that burrow through your body, eat your organs, and kill you. There are amoebas that can get behind your contact lens, eat your eyeball, and make you go blind. There's even something called a pinworm that lives in your rectum until it's ready to lay eggs, then waits until you are asleep, sneaks out your anus, and lays eggs so you'll be itchy in the morning, touch them with your fingers, and accidentally pass them on to other people when you cook breakfast.[12]

Another noteworthy human parasite is the leech, a blood-sucking animal closely related to the earthworm. A leech has two suckers—one at the mouth and one on the tail—and it uses them in synchrony, to move around like an inchworm. I remember a fellow bat researcher once telling me about the time he was on a bat expedition in the forests of Vietnam and took a moment to squat in some tall grass to poop. As he squatted, the tips of the grass were at eye level, and he could see leeches on those tips, extended upward from their bottom suckers, waving like alien finger puppets all around him. Every time he rustled to adjust his position, the leeches would bend over and inch their way toward him, then pause once more when he got still again.[13] Despite his best efforts at stillness, he told me, he had some removals to perform after the job was done. I didn't ask him for details.

Leeches feed on people, but they feed on all kinds of other animals too. Some may simply attach themselves to an animal's leg, but there are other species that specialize in swimming up an animal's nose when it takes a drink from a stream. Then it painlessly attaches itself to the host and starts drinking.* Once a leech is attached, it sucks blood until it swells to around ten

* I say "painlessly" because the show about parasites I work on, *Monsters Inside Me,* once featured an interview with a guy who got a leech up his nose in Nepal. He didn't even know it was there until he started getting nosebleeds, and then later periodically saw it stretching out of his nose, waving in front of his face. Throughout his experience he felt no pain, so I assume the animals upon which these leeches typically feed also feel no pain.

times its original weight. Then the next time the animal takes a drink from a stream, the leech lets go, floats away, and digests quietly for a few months, leaving its host only a little worse for wear.

Typically, humans do everything they can to prevent being fed upon by a leech, but that's not always the case. There are instances where a leech can be a huge help to a person. For instance, let's say a great white shark bites a few of your fingers off, or maybe a dog gets its lower jaw inside your mouth when it's attacking you and tears your whole cheek off, or maybe you just lose part of your face to cancer (heaven forbid). In those cases—which I did not make up, by the way—surgeons can reattach the fingers, cheek, or grafted piece of flesh from your own forearm, and they use leeches to help them do that.[14]

You can't just sew a piece of flesh back into place—you need to make sure there's a constant flow of blood into and out of the attached flesh, or the cells within it will die. Surgeons reattaching a flap of skin need to make sure arteries from the body pump blood into the flap, and that the veins within that flap link properly to veins within the body to drain blood out again. But lining up blood vessels is hard, and sometimes blood starts to build up in little pockets when the draining isn't working quite perfectly. Letting that blood pool in the tissues is bad, but by putting leeches on the surface of the reattached flesh, surgeons ensure that blood doesn't accumulate anywhere. Leeches just suck blood out of the area so that it can't accumulate. So long as the leeches are switched out frequently for new ones, the whole blood network of the region has time to heal.

There's something wonderful to me about the fact that leeches do a better job draining human blood from a surgical site than even our most sophisticated technologies can. The sucker of a leech has over one hundred teeth to make a clean cut, but the leech provides its own anesthetic to make the bite less painful, and its own anticoagulant to keep the blood flowing. Best of all, leeches don't leave scars.

It's worth mentioning that when humans take advantage of leeches this way, the leeches don't count as parasites anymore. Remember, a parasite has to impose an overall cost on its host. Having your cheek heal when it's reattached is a benefit, so leeches used in surgery don't count as parasites, even though leeches of the same species would count as parasites if I found them on my leg. Parasitism is all about the relationship. In a sense, we're taking slothful animals that would rather spend their time mooching, and putting them to work for us.

That's not the only job we've given to leeches either. Researchers have started using leeches to help with the conservation of rare mammals in tropical countries as well. Instead of tromping through the forests of Vietnam looking for animals so rare you might go months without seeing one, researchers can go to a patch of habitat, catch leeches there, then check the DNA of the blood inside the leeches to find out what animals they've been feeding on. A leech can store blood for months after a meal, so a single day sampling leeches from an area can give hints about all the different animals that have been there in the last few months.[15] Using this method, researchers have found the DNA of extremely rare Vietnamese mammals, including the Annamite striped rabbit, the small-toothed ferret badger, the Truong Son muntjac deer, and a strange antelope called a serow. You might never have heard of any of those animals (don't feel bad—I hadn't either), but that's the point. Even conservation biologists working in those very countries know almost nothing about those creatures, and leeches have become one of their few sources of information about them.

There are a lot of different kinds of parasites, but sometimes what looks like a parasite can turn out to be something else entirely. There's a fish called Krøyer's deep sea anglerfish that lives in the pitch dark, one or two kilometers below the surface of the ocean. The female is about two feet long, with a long, thin pole sticking out of her forehead. At the tip of that pole, she has a sack containing bacteria. Those bacteria produce light, and the light they produce attracts smaller fish, which the anglerfish then eats.[16] You might remember seeing a deep sea anglerfish in the movie *Finding Nemo*, when she lured Marlin and Dory in with her light organ and then almost ate them.[17]

Since the bacteria provide a benefit to the anglerfish, you know they're obviously not parasites, but that's not where I'm going with this. If you look carefully at a female Krøyer's deep sea anglerfish, you'll usually find something sticking out of her belly that looks a lot like a leech. It's not a leech, though. It's another fish. At first, that smaller fish was assumed to be a parasite, but it turned out to be a *male* Krøyer's deep sea anglerfish.[18]

Because males and females of the species meet only rarely down there in the dark, they use a strategy that will ensure that the male will be nearby when the female is ready to lay eggs. When they meet, he burrows his mouth into the bottom of her body and then stays there. Eventually her skin grows around him, and they become one self-fertilizing hermaphroditic Franken-fish. Or if you prefer, she becomes a female with a scrotum. However you want to describe the arrangement, she does all the hunting for both of them: she has a bioluminescent lure and a giant mouth, and while he has neither of those things, he feeds on her body fluids, which *almost* makes him a parasite. But since he makes sperm for her, he's not a parasite. He's a pain, for sure, but she gets a benefit from the relationship, so you can't call him a parasite no matter how tempting it is to do so.[19]

Anglerfish use a parasite-like lifestyle to cope with the difficulty of finding members of their own species in the abyss of the deep ocean, but it could be worse. If you think it's hard to find a sex partner in the deep sea, imagine being a parasite inside the body of your host and then having to search for a mate. You could search forever, and there may simply not be any other parasites of your own species in there to find. That's the potential predicament faced by a tiny flatworm that sometimes infects people. And that's why when a couple of those flatworms finally do hook up inside a person, they make that romance last.

This particular flatworm, called a schistosome, starts out inside a snail.[20] When the parasite first swims out of the snail, it has just eight hours to find a human host before it dies. If it's lucky enough to find someone, it burrows right through the person's skin, enters the bloodstream, and then floats to the lungs. There the worm rests for a couple of days, builds itself a jacket of proteins so the human's immune system won't recognize it, then slips back into the bloodstream. Eventually, it finds its way to the small blood vessels near the human's bladder or intestines. This is where the search for a mate must happen, and as you can imagine, the odds of finding a mate there may be very, very low.

If a male and female schistosome do happen to cross paths

there, the smaller male curls up inside a groove that runs down the body of the female.* He's like a half-inch-long hot dog, and she's a slightly longer hot-dog bun. With things in place, they start having hot-dog-on-hot-dog-bun sex, and that sex lasts for up to thirty years.

Over that time, they will release roughly three hundred eggs per day into the blood. (Thirty years of that results in more than 3 million eggs.) Half those eggs will be absorbed by the human host, causing all kinds of problems, like blood in the feces and urine, for example, and half will make their way into the host's urine or feces, then out of the body.[21] With any luck (for the parasite), those wastes will get back into the water so that some other snail can be infected, and the life cycle will continue.

With the odds so low that a male and female will ever happen to meet in the same blood vessel at the same time, these flatworms have adopted a strategy of just making zillions and zillions of eggs in the hopes that a few of them have that kind of luck. The good news for schistosomes is that it seems to be working. Two hundred million people have this disease, which is called bilharzia or schistosomiasis. But of course, what's good news for schistosomes is bad news for humans.

Unlike malaria, which can kill you, bilharzia just wears down the immune system. That makes the people who suffer from it susceptible to other diseases or infections. Fear of bilharzia once prevented me from dipping my feet in a cool river after a long hot day of bat work in Madagascar in 2008.† I could see the locals washing themselves and their clothes in the river, but my knowledge of

* In fact, the Greek word *schistos,* which gives schistosomes their name, translates as "split" or "divided."

† It was in the tiny village of Kianjavato, and we were there searching for the elusive Madagascar sucker-footed bats.

parasites made that pristine stream look like the jaws of death. I was taking drugs at the time to prevent me from catching malaria, but I was defenseless against the schistosomes. They terrified me.

You might have had experience with schistosomes too, even in Europe or North America. If you've ever had "swimmer's itch," your skin was penetrated by schistosomes that normally infect birds.[22] Fortunately, they were killed by your immune system before they could do any damage. Those itchy welts on your legs, which resulted from the histamines your immune system used in its assault against the worms, serve as a good reminder that even though most parasites can't infect humans, they truly are everywhere.

Many of the parasites that infect nonhuman species mirror our own, like bird schistosomiasis or gorilla malaria, both caused by parasite species slightly different from the ones that infect us, but many animal parasites are totally different from anything humans ever get (thank goodness). The slothful but successful strategies of parasites have resulted in some breathtakingly disgusting spectacles.

One of my favorites (I'm literally wincing as I type this) is a wasp, called the emerald cockroach wasp, that will fly out of the sky and sting a cockroach, so that the cockroach, though still able to walk, just really doesn't want to anymore. Next the wasp grabs the roach by an antenna and drags it, kind of like a poorly trained dog on a leash, to its underground lair.[23] Then the wasp lays an egg on the cockroach, buries it, and leaves. When the egg hatches, the maggot burrows into the cockroach and begins feeding. To keep the roach alive as long as possible, the maggot eats the roach's organs in a careful sequence. It also spits cleaning chemicals all over the place to make sure things stay fresh—that is, to make sure bacteria don't grow in the insect, which would make the maggot's food source go bad.[24] These strategies allow the maggot to feed on that same defenseless cockroach for several weeks, and what's really sick about it all is that the cockroach stays alive right up until the end, when the maggot molts into an adult wasp, bursts out, and then tunnels its way up through the soil to the surface.

You can hate cockroaches as much as you want, but you have to admit that's a pretty terrible way to die.

The thing to keep in mind, though, is that this happens *every time* an emerald cockroach wasp reproduces. They can only reproduce by torturing cockroaches. It's not like a wasp here or there suddenly goes all evil and starts needlessly torturing other living things. Every time an emerald cockroach wasp hatches, it's from the corpse of a tortured animal.

This parasitism is so ruthless that the animals that lay their eggs inside a living host have been given their own special name. They're not just parasites. They're parasit*oids*. In fact, there are many wasps, flies, beetles, moths, and other insects that make their living this way. Believe it or not, 10 percent of all insect species are parasitoids. If you just look at the flies, you can find 16,000 parasitoid species—roughly a fifth of all the fly species in the world![25] In other words, laying eggs in a living host isn't some biological rarity. It's common. And that means millions and millions and millions of animals around the world are being tortured this way as you read this sentence.

There are many counterstrategies for the would-be victims of parasitoids. Some fight off the egg-laying attacker, whereas others just try to avoid them in the first place. Because parasitoids sometimes track down a host by the smell of its poop, one way to stay hidden is to make sure you're never very close to your own feces. That's why the skipper caterpillar of Brazil launches feces at around 4.3 feet per second whenever it goes to the bathroom, propelling its poop up to 30 inches away.[26] To put that in human terms, that's a five-foot-tall woman launching her poop 75 feet (from a lying-down position, no less).[27]

For many years it was assumed that this fecal feat was accomplished by means of a flicking set of brushes, called an anal comb, that could launch the mess, much like your toothbrush can spatter toothpaste all over the mirror if you rub the bristles. But no, nature is far grosser than that. We now know that the fecal flinging is accomplished through a buildup of pressure, just as you would launch a spitball through a straw. I've had some pretty explosive diarrhea in my day, but even amid the gastrointestinal fireworks I experienced in Madagascar, I haven't come close to the performance of a skipper caterpillar.

The only reason this super pooper ever evolved in the first place is the deadly threat of parasites. Parasites are important that way: they shape animal bodies, animal behavior, and animal evolution. When people describe evolution, they usually talk about how hard animals have to work to avoid getting eaten by predators, but parasites may well be even more important in most circumstances.

Sometimes, even when it looks like predators are picking who lives or dies, it can be parasites who are really in control. For example, one particular bat in a barn will get eaten by a raccoon, as opposed to the other bats around it, but that may be because that particular bat was sick from its parasites. For the parasites inside that prey animal, getting eaten could mean they'll get into the predator. Suddenly sickness isn't just a failure of the prey animal's meat robot but the hijacking of that meat robot by the parasite's DNA. It's enough to make you look at sickness and health in a whole new light.[28]

If that weren't enough, sometimes parasites take things even a step further, changing the behavior of their hosts in surprising ways to improve the parasite DNA's odds of survival. There are many, many examples of this, but my favorite mind-controlling parasite is a creature called *Toxoplasma*.

Toxoplasma is a parasite of cats, but it also spends some of its life cycle inside a rat, and the rats are the ones who have their minds controlled. Here's how the life cycle works. First, a rat accidentally eats *Toxoplasma* eggs when it eats something contaminated with cat feces. In a few weeks, those eggs turn into cysts all over the inside of the rat's body. When that infected rat gets eaten by a cat, those *Toxoplasma* cysts get into the cat's stomach. Later, the cat poops, and that poop gets onto something the rat eats, and the life cycle repeats. Simple, right? The tricky part of this life cycle (for the parasite) is finding a way to cause the rat to be eaten by a cat. Apparently, that's much harder for the *Toxoplasma* parasite than you might expect.[29]

The problem (for the parasite) is that rats are really good at avoiding cats. For example, if a rat ever comes across the smell of cat urine, nerve signals from the nose go straight to the fear centers in the rat's brain, and the rat instinctively gets out of there. In other words, cat urine smells terrifying to a rat, and that's hardwired into the rats by their DNA. Even rats raised in captivity are scared of the smell of cat urine the very first time they smell it. Those rat instincts are a real problem for the parasite, which needs its host rat to be eaten by a cat. Otherwise, the parasite's DNA will never be passed on.

Here's *Toxoplasma*'s solution: When the parasite encysts in the rat, many of those cysts form in the rat's brain, and some of those form right at the very fear centers normally triggered by the smell of cat urine. Somehow, incredibly, those cysts alter the wiring of the brain so that when a *Toxoplasma*-infected rat smells cat urine, the nerve signals don't go to the fear centers at all but instead get rerouted into the sexual pleasure circuitry of the brain. In other words, rats infected with *Toxoplasma* don't just lose their fear of cat urine, they get sexually aroused by it.

Not surprisingly, this makes them more likely to spend time near cats than normal rats would, and that improves the odds that the parasites will complete their life cycle.

This is what I'm talking about when I say parasites are just as important to ecosystems as predators and prey. You can watch a cat eat the rat, but if you leave out the fact that the parasite is controlling the rat, you can't really ever understand what's going on.

Take a second to imagine that: What if there were some parasite that could alter our own behavior in that fundamental a way? If that idea freaks you out, you may just want to stop reading here. If the thought of being a puppet controlled by a microscopic parasite lodged in your brain is just too creepy for you to handle, maybe this isn't the book for you.

Still with me?

Here it comes: Sometimes humans accidentally eat cat feces, just like rats do. When that happens, humans can end up with *Toxoplasma* cysts in their bodies, just like rats do. In fact, that happens all the time. By some estimates, a third of the world's population is infected. *A third!* I'm not just talking about developing countries either. The infection rate in the USA is around one in eight people, and in some countries that number is closer to seven out of every ten people. That might be shocking, but it makes sense: cats (and rats) live pretty much everywhere that humans do. Changing the kitty litter every few days for several years, you're bound to have a little fecal dust go into your mouth at some point.[30]

Toxoplasma, by the way, is the reason women who are pregnant aren't supposed to clean up kitty litter: the litter might be filled with *Toxoplasma* eggs, and although the parasite is relatively harmless to adult humans, becoming infected while you're pregnant can cause severe damage to your unborn baby.[31]

At this stage you probably have a lot of questions. Do you have *Toxoplasma* cysts in your tissues? You very well might. And if so, will the smell of cat urine make you horny?

The answer to the horniness question is probably no. It doesn't look like the parasite affects people the same way it affects rats. Rewiring does happen, but it doesn't cause us to become sexually attracted to the smell of cat urine.* It does other things instead.

For one thing, the reaction time of a human infected with *Toxoplasma* is around twelve milliseconds slower than normal. That's a small difference, but it has been assumed to be the reason people with *Toxoplasma* infections are involved in car accidents more frequently than uninfected people are. Also, when scientists give standardized personality questionnaires to people, those infected with *Toxoplasma* score differently from people without those infections.[32] However, the changes are subtle and hard to interpret. Men seem to disregard rules and act more suspicious and jealous, whereas women seem to be more warmhearted and easygoing. Both sexes show a decreased urge to try new activities. It's all quite confusing, as though the parasites are tinkering with our minds but haven't really decided what they want to change about our behavior.

* If cat urine does turn you on, I'm afraid you'll need to find a different excuse. Sorry.

In fact, that's probably exactly what's happened. A human brain and a rat brain are similar, but they're sufficiently different that the parasite can't use its rat-brain strategy in a human. Instead, *Toxoplasma* is following its DNA's instructions, making nonsensical adjustments to its human host's brain that would have worked in a rat. Since none of the parasites that end up in humans will ever get inside a cat, natural selection hasn't been able to fine-tune those manipulations into anything meaningful.

Even though those changes to human behavior don't serve any useful function for the parasites, they certainly make a difference to the humans. For one thing, people have suggested that *Toxoplasma* could be responsible for many of the cultural differences that exist between populations in different parts of the world. For instance, some might say that "Men disregard rules and women are warmhearted" is more true of Brazil, where the *Toxoplasma* infection rate in humans is 67 percent, as compared to South Korea, which has a 4.3 percent infection rate. People love to make jokes about how cats manipulate their owners, but when you factor in cat parasites, those jokes are suddenly way less funny.[33]

When *Toxoplasma* infects its intended host, the rat, it manipulates that host in a way that will cause the rat to die, because that manipulation will help the parasite live on. But sometimes a parasite's best strategy is for its host to stay alive. In those cases, a parasite will take care of its host, and my favorite example of this is a story that involves bats. It's a great story. I told it to my dad more than ten years ago and he still brings it up from time to time.

The parasite in question is a mite that lives on the ear of a moth. That moth uses its ears to listen for the calls of approaching bats. That's an important thing for a moth to listen for, since bats hunt for moths using echolocation—shouting at high-pitched frequencies in the dark. When the moth hears a bat coming, it can do an evasive maneuver. A moth not listening for bats in the night would be as vulnerable as a gazelle in the daytime wandering blindfolded among lions.

Those ears, by the way, are located on the wings of the moths and are tuned to the high-frequency echolocation calls of bats—sounds so high-pitched that no human could ever hear them. There's one ear on each wing and, like your ears, they operate independently of each other.[34]

The problem is that when parasites inhabit a moth's ear (and there's nowhere else they can live, by the way), they lay hundreds of eggs—and cause the ear to stop working in the process. That's bad for moths because it means they can't hear bats coming to eat them, and that, in turn, is bad for the parasite because once its moth gets eaten, the parasite dies too, along with all her eggs.

Now listen up, because this is the part my dad still can't get over. A colony of mites will only ever infect one ear of a moth. Sometimes it's the right and sometimes it's the left, but once one ear is infected, mites will never infect the other ear. Why? Because that way the parasite gets the best of both worlds. The moth can still hear bats coming with its one good ear, and the mites get to raise their offspring. The mites have to hurt their host to live, but they take good enough care of that host to keep it alive.[35]

The parasites are kicking butt out there, lazily stealing from the hard workers, and sometimes even making them do their bidding with mind control. Sloth is alive and well in nature. From vampire bats to *Toxoplasma*, the parasites are making nature more complex, more interesting, and (I would argue) more beautiful than it would be without them. And like the anglerfish that borrows from the parasite playbook during reproduction, we humans pay homage to parasites early in life as well.

Surely it's occurred to you that a human fetus acts a little bit like a parasite to its mother, hasn't it? As a fetus, Sam took oxygen and nutrients right out of Shelby's bloodstream, and Sam's fetal body even released chemicals that suppressed Shelby's immune system so her body wouldn't reject him. If imitation is the sincerest form of flattery, we humans make overtures to the parasites with every passing generation.

A mother and her fetus are very separate organisms, and sharing a body for nine months is hard on both of them. For example, the tension between a maternal immune system and the fetal immune system can have some surprising consequences having to do with homosexuality. The mother's immune system develops a stronger and stronger immune response to the male antibodies produced by each successive son. The fact that younger sons get attacked more aggressively by their mom's immune system is thought to be the reason that those younger sons are more likely to grow up to become homosexual. To be clear, this isn't the reason homosexuality exists, but the mother's immune response is probably the reason that the likelihood a boy will be a homosexual increases by 33 percent for each older brother he has. In the first son, it's around 2.0 percent. In the second it's 2.7 percent, and so on, to as high as 8.5 percent in the sixth son. This correlation only seems to apply to male homosexuality, not female, and the number of female siblings has no influence.[36]

And just as the fetus must survive the onslaughts of the maternal immune system, the mother must do that and more. However, despite the morning sickness and the stretch marks, and all the other costs during pregnancy, not to mention the ordeal of childbirth or the considerable work it takes to look after a human being, a child cannot ever be a parasite to his mother. A mother gets such an enormous benefit from her baby's existence that it outweighs any and all of those costs. It's the benefit Shelby gets from Sam, and the same benefit I get from Sam, as his father. Through Sam, our respective, separate DNA sequences have been passed on so that they can survive, even after their meat robots die. In the game of life, there's no greater prize than that.

Parasites complicate the meat robot story, though, because parasites turn the very concept of individuality on its head. When a rat gets sexually aroused by cat urine, is the rat's body still acting as the rat's meat robot, or has the rat's body now become the meat robot of *Toxoplasma*? When a human spills hundreds of millions of schistosome eggs into a stream through his urine, is that human working for his own DNA or for the DNA of the schistosomes inside him? When a moth survives a bat attack, does it have its own DNA to thank or the DNA of the mite? When Sam wakes up at 3 a.m. for no reason at all, and I run to his room to rock him back to sleep, am I my own DNA's meat robot, or has Sam's DNA been pulling my puppet strings? All of a sudden, the DNA strands piloting the meat robots of the world become less discrete. It's the ultimate in sloth—if you're too lazy to build your own meat robot, find a way to hijack someone else's.

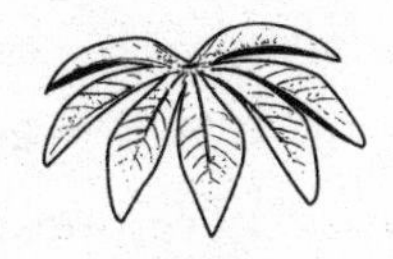

4

GLUTTONY

Eat and Be Eaten

One of the classic arguments people use to try to prove that nature's actually really nice to us is the wealth of foods she provides. You can load up your grocery cart with fruits, vegetables, breads, cheeses, eggs, meats, and all of them came from the natural world. Even the high-fructose corn syrup in your gummy bears came from a plant. It's a fair question: If Mother Nature's so selfish and violent, why does she do such a wonderful job of keeping us alive?

That supposed care she provides is an illusion. She's not really trying to feed us at all. Instead, we've just become masters of working nature over to feed ourselves. Plants and animals weren't put here to take care of us. They appeared as the result of evolution, just as we did. They are other kinds of meat robots doing their best to survive and reproduce, just like the rest of us. (I suppose it's a little strange to call a plant a "meat robot," but I'm going to do so anyway, just to be consistent about the relationship between DNA and the body it makes.)

Plants are selfish, so they go to extraordinary lengths to *avoid* getting eaten. They produce everything from thorns to poisons to keep animals away. In response, we've learned to avoid most of the plants altogether and focus instead on the small minority of plants that we can eat. We actually don't eat very many kinds of plants at all. We've selected a very small number of plants that don't hurt us and then bred them to make them even more delicious.[1] For instance, cabbage, kale, broccoli, cauliflower, brussels sprouts, and kohlrabi all come from the same species of plant, *Brassica oleracea*. The plants aren't trying to help us; we're helping ourselves.

As for the animals we eat, I think it's a safe bet that they don't really want to be eaten by you or have their unborn babies made into omelets or have the milk they make for their own babies mixed with your coffee. Fish may be a great source of omega-3 fatty acids, but those molecules aren't inside that fish for your health. They're there because they're good for the fish. When you go for sushi, you're stealing parts from that fish's meat robot and incorporating them into your own.

Plants and animals make us healthy because our bodies have evolved in direct response to the plants and animals we eat. We've been killing and eating nature's bounty for millennia, so human bodies are built to thrive on precisely those kinds of foods. To say nature is there to make us healthy has it backward. Our bodies are built to steal from the other bodies around us: we can only survive by eating plants, animals, and fungi.

But we're not the only ones out there with an appetite. To

get a clear perspective on our own gluttony, it helps to first see how intense gluttony can get in nature. That's what this chapter is about—gluttony in the natural world—starting with plants that make their own food, moving up through the plant-eating animals, the animal-eating animals, and finally to the animals that dine on the corpses that are left behind in the carnage. Gluttony abounds.

\\\\\\

There's a man in India named Prahlad Jani who has consumed neither food nor water since he was eleven years old, in 1940. How do we know this? Because he says so. Mr. Jani says he gets all his energy from the sun.

That seems like it should be impossible, but there's a physician in India, a *neurologist* in fact, named Dr. Sudhir Shah, who has verified Mr. Jani's claim on two separate occasions. Over the course of ten days in 2003 and fourteen days in 2010, Dr. Shah and his team watched Mr. Jani carefully and confirmed that he survived with no food or water.

Well . . . no water *except* the water he got each day to rinse out his mouth, but he promised not to swallow any of that. Also, he got to bathe himself, but other than those two minor details, he had absolutely no food or water at all over the course of those two studies.[2]

Now, for reasons you might be able to guess, neither of Dr. Shah's studies has quite made it into a scientific journal, but Dr. Shah has published a PDF on his website, explaining how Mr. Jani might have turned himself into a "kind of solar cooker" with "solar batteries."[3] This story has been covered by newspapers and

TV programs around the world, and some people actually believe it's true.

Of course it's not, though.

I don't know if Mr. Jani is a fraud or if he is suffering from a mental illness that makes him unaware that he eats and drinks, but I do know that humans use energy to stay alive, and I also know there's no way for a human to get energy from a beam of sunlight.

As for Dr. Shah, he's either in on the lie or he's not a very good doctor (or both, I suppose). A human cannot go decades without food or water. That seems like the kind of basic fact about humans that a doctor should know. I don't think I'm being unfair about this.

Humans need water. A person with no access to water at all may die in less than a week. Our bodies are mostly made of water, not 99 percent as the urban myth goes, but closer to 60 percent (the proportion changes slightly with health and age).[4] And that water is constantly streaming out of us: as urine, feces, sweat, tears, the humidity of our exhaled breaths, and for women, during menstruation. With so much water leaving the body, our food and drinks need to bring in at least 2 to 3 liters of water per day (that number can vary, but 2.6 liters is the number scientists have used when trying to estimate how much water astronauts will need).[5] That's about five and a half pounds. During heavy exercise or in hot climates, those water needs can more than double.*

* It's surprisingly hard to find an answer to the question "How long can a healthy person survive if they stop drinking water?" The rate at which people lose water will depend on temperature, humidity, amount of exercise they're doing, and their overall health. Ellershaw et al. (1995) showed that elderly people with malignant disease who stopped drinking water died within one to five days. A young, healthy person could live longer than that, but likely not much more than a week.

In addition to water, humans also need food, because that's where we get our energy. A person on a hunger strike (drinking only water) usually dies after a month or two, having burned up all their energy reserves. Even just sitting around, a person burns through the equivalent of around 580 AA batteries each day.[6] With daily activities factored in, like walking, talking, working, and all the other things a person does, daily energy costs can be two to three times higher than that. It's simple physics. Energy cannot be created or destroyed. Since Mr. Jani's body uses energy, he has to consume energy.

Mr. Jani's claim is that he gets his energy from sunlight. Plants can do that, but animals like Mr. Jani cannot. The process by which plants do that is called photosynthesis.* Photosynthesis first evolved about 2.4 billion years ago, so long ago that everything living on Earth still had only one cell and lived in the water.[7] Those first photosynthetic organisms therefore weren't plants as we know them today but single-celled algae, like those that make up the mat you might see on the surface of a polluted pond, for example. Many kinds of single-celled algae are still around today, but one lineage of algae has changed a lot since then, becoming multicellular, adapting to life on land, and becoming the plants we know and love. (In other words, plants are specialized kinds of algae.)

Plants and algae can photosynthesize because they're direct descendants of those first photosynthetic organisms and have inherited the complex machinery required to harness the sun's energy. Animals can't photosynthesize because they don't have the

---

* *Photo* means "light," and *synthesis* describes the sugars that plants make (i.e., synthesize) with the solar energy they gather.

equipment. The idea that Prahlad Jani suddenly just happened to be able to make sugars from sunlight would be like discovering one day that a brick factory could suddenly make sports cars.

You can't harness the power of the sun by closing your eyes and having warm thoughts. It's a precise chemical process. Photosynthesis requires dozens of incredibly specialized proteins that humans don't have, all working together in perfect synchrony, like robots on a microscopic assembly line. A beam of sunlight penetrates the surface of a leaf, exciting a special molecule inside it, called chlorophyll a. Left alone, that excited molecule would give off some of that energy as light, glowing red as the energy dissipated. But in the leaf, it doesn't glow, because crowds of orderly proteins around that molecule immediately jump into action, harnessing its energy to rip molecules of water (H_2O) and carbon dioxide (CO_2) apart, then rebuild their constituent atoms to make sugars ($C_6H_{12}O_6$).

Oxygen (O_2) molecules, by the way, are released from this process as a waste by-product, the result of the plant's having extra O_2 left over from the breakdown of the H_2O and CO_2. But one life form's trash is another life form's treasure: oxygen is a waste product we've built our lives on. (More about that in the chapter on wrath.)

Have you ever played with those halves of hollow rubber balls that you can flip inside out and then place on a table, until a few seconds later they spontaneously pop back into their previous shape and go flying into the air? That's how I like to think of sugars. It takes energy to bend a half ball into that inverted shape, and that energy is physically stored within the structure of the half ball itself. When the half ball pops, the energy is released as the object relaxes to a more "comfortable" shape. Sugars basically work just like that. It takes energy for a plant to load atoms into a sugar, and that energy can sit there, inside the sugar, to be released later on. When a sugar is broken back down to water and carbon dioxide, the energy comes popping out. Plants build sugars so they can use the sun's energy later on—for growth, reproduction, and whatever other processes are required for staying alive. But if an animal steals a sugar by eating part of a plant, the animal can break the sugar down itself and use the energy for its own purposes.

As you read this book, your eyeballs move from side to side because they're pulled by muscles. The energy burned by those muscles originally came from plants. It's kind of a shame, when you think about it. All this energy rains down on us from the sun each day, but as animals we can't tap into any of it. Instead, we let plants do that work, and then we eat the plants. It's as if instead of bringing lunch money to school, we beat up the kids who bring their own lunch each day, and take theirs.

Whole ecosystems work this way. The energy inside a deer came from solar-powered plants. When a cougar kills a deer, the plant energy is passed on—to the cougar, but also to the cascade of small mammals, birds, insects, fungi, and bacteria that clean up the bits of carcass left behind by the cougar. All those living things (and their parasites, by the way) constantly fight one another to get energy into their bodies. Then the game becomes one of making sure no other animals steal the energy from them. Energy is constantly flowing through nature, and gluttony is the mechanism by which it does so.*

* I should note that photosynthesis isn't the only route for energy to enter our living world. There are some bacteria that can harness energy from the breakdown of chemicals like methane or hydrogen sulfide. But they only live in extreme environments, like

\\\\\

As the nozzle through which energy flows into ecosystems, plants have it pretty rough. There's an endless parade of animals, from aphids to zebras, constantly trying to take a bite out of them. Plants can't run and hide, so they have to hunker down and defend themselves. As a result, plants make up some of the most violent, ruthless, and lethal instruments you'll ever find in nature's arsenal.

It's easy to think of plants as harmless, as we look around the produce section of the grocery store, but of course that's because none of the dangerous plants are there. In the past, foraging for fruits and vegetables meant hunting among hundreds of inedible plant species for something edible. There are more than 250,000 kinds of plants in the world, but we humans get more than 90 percent of our calories from just fifteen of them. That's around a 160th of 1 percent.* Most of those plants have been cultivated by humans for thousands of years to make them better for us than nature originally made them, or at least more appealing or easier to eat. Grocery store advertisers may use the word *natural* to describe their foods, but the produce section of the grocery store is a far cry from walking across the African savannah looking for something to eat.

To defend themselves, many plants use thorns and spines. (Ever tried to chew on a rosebush?) Sharp, poking bits can make

---

hydrothermal vents at the bottom of the ocean, and the amount of energy they bring into the world is minuscule compared to what comes from the sun. So for simplicity I'm going to focus only on the energy that comes into ecosystems from photosynthesis.

* Those fifteen plant species include things like wheat, rice, corn, potatoes, beans, and so on. If you make a list of all the plants humans have ever eaten even a tiny piece of, that number jumps to around one-eighth of all plant species (Pimentel and Wilson 2010).

it painful to touch a plant—never mind eat one—and plants often make those spikes even more effective by filling them with noxious chemicals. That way, animals that try to eat the plant get blisters and sores to remind them not to do that again. However, by far, the most impressive set of spines on any plant has to belong to the bull's horn acacia. Working on the principle that sometimes the best defense is a good offense, that plant has found a way to keep weapons inside the thorns that can crawl out to repeatedly sting anyone who comes too close. Even more incredibly, it does this by secreting a chemical that isn't noxious at all. The acacia plant secretes a nectar.

That nectar is there to feed tiny, vicious, wasplike stinging ants that make their home inside the plant's hollow branches and thorns. The ants live nowhere else—that's why they're called acacia ants. The ants don't hurt the plant, but they get all their food from it. Acting selfishly, the ants defend that food source against other animals, and that works out very nicely for the plant. If some deer takes a bite of bull's horn acacia, it ends up with a mouthful of stinging ants.[8] And these ants' stings are particularly painful (a "rare, piercing, elevated sort of pain," as you will see in the chapter on wrath). Acacia ants can even turn elephants away.[9]

The ants aren't parasites of the plant, since the plant gets a benefit. In biological terms, their relationship is called a mutualism: both parties benefit. The plant gets protection, and the ants get room and board. But it's not all hugs and kisses between them. The two species have been working together so long that the ants no longer have the ability to find food anywhere else, and that gives the plant a lot of control in the relationship. For example, the amount of nectar the plant is willing to produce varies depending on how worried it is about herbivores. Over the course of the year, and even throughout the day, the plant gives up as little sugar as possible, only secreting more when it needs the ants to provide better protection. The number of ants that can live on the plant's nectar rises and falls at the whim of the plant. You can think of the plant as a millionaire, paying a fleet of security guards as little as possible to defend its factory, with some dying in the fight against intruders, and others simply dying of starvation when layoffs happen during periods of low crime.

The bull's horn acacia has turned acacia ants into a living, swarming defense weapon, and it decides when and where that weapon will be ready for use. In essence, the plant has enslaved the ant meat robots to work for its own DNA. From the plant's perspective, the ants have pretty much become part of its own body.

Although enslaving ants is effective as a defense system, it's a solution to herbivores that hasn't evolved in most plants. Instead, the vast majority of plants do something simpler to deal with herbivores: they just make themselves poisonous. It's estimated that plants have come up with more than 200,000 different chemical compounds, and the effects those chemicals have on the animals that eat them can be wonderfully brutal.[10] One challenge with poisons, though, is that the poisons most lethal to animals are often also deadly to plants. That's bad news for the plant. There's no point in making poisons to keep animals from eating you if you're going to kill yourself in the process.

For example, hydrogen cyanide is extremely lethal to animals. The lethal dose for a human is around half a milligram of cyanide per pound, so one hundred milligrams could take out a two-hundred-pound person. (For reference, a toothpick weighs around one hundred milligrams.)[11] Hydrogen cyanide kills animals because it interferes with the molecular machinery used to get energy out of sugars. Since plants also get energy from sugars, hydrogen cyanide kills plants too. But despite that, more than 2,500 species of plants produce cyanide to deter herbivores.[12] Somehow, though, the plants don't succumb to their own poison, and the way they achieve that is brilliant: they basically pack hydrogen cyanide into bombs that will only go off if the plant gets eaten.

Here's how the bombs work: Instead of making hydrogen cyanide ahead of time, the plant builds a bigger molecule that has hydrogen cyanide inside it. Because the hydrogen cyanide is stuck inside that larger molecule, it can't perform the chemical reactions that would otherwise make it poisonous. That part is the bomb. Simultaneously, the plant builds an enzyme that can break that hydrogen cyanide free from the rest of that molecule. You can think of that enzyme as the bomb's detonator. The plant stores the bombs in tiny walled-off sacs throughout its tissues and then surrounds those sacs with clusters of detonators. So long as the plant is uninjured, the chemicals stay separate, and no poison is ever made. But once a herbivore takes a bite of the plant, the sacs are mechanically broken by the predator, the detonators set off the bombs, and the lethal hydrogen cyanide is secreted right into the mouth of the herbivore. It's a perfect system, and the real beauty of it is that the uninjured parts of the plant don't get poisoned at all. Only those parts of the plant that are already being chewed on end up getting sacrificed.[13]

Not surprisingly, hydrogen cyanide creates problems for a lot of herbivores. Gorillas and rhinos, for example, eat plants that defend themselves with hydrogen cyanide, but they limit those plants to a small fraction of all the different foods they eat, presumably so that the poison dose stays small enough not to hurt them.[14]

Anyone who's watched a James Bond movie or seen an Agatha Christie play knows that cyanide is deadly to humans, but humans routinely eat plants that produce hydrogen cyanide. Cassava (also called manioc) is a root, sort of like a potato, that is a staple in the diet of roughly 500 million people worldwide. It's mostly eaten in Africa, the Philippines, and Brazil but also makes its way into kitchens in Europe and North America.[15] The roots of that plant are deadly if eaten right out of the ground, which is part of the reason it works well as a crop: few animals can spoil the harvest by eating your crop beforehand. Humans get past the hydrogen cyanide in cassava by presoaking, fermenting, or cooking it, to break down the dangerous chemicals inside it. From time to time, though, people do die as a result of eating unprocessed cassava. It's a sobering reminder that we survive by exploiting the plants and animals around us. Mother Nature isn't trying to keep us healthy. We're just taking what's available in nature to look after ourselves.

The other strategy plants use to avoid self-poisoning is to make poisons that hurt animals but have no effect on plants. That way, the plant can produce those chemicals to its heart's content without ever having to worry about getting hurt. Such chemicals usually work by focusing on animal body parts, like nerves, that plants don't have. One plant, called zonal geranium, puts a drug in the petals of its flowers called quisqualic acid. Beetles that eat the petals feel fine at first but after about thirty minutes start to realize their hind legs don't work so well, and before long they are totally unable to move. From tests in the lab, scientists know that the drug's effects last only about twenty-four hours, but out in the woods where this plant grows, a beetle lying defenselessly on the ground for a day is almost always eaten long before its paralysis is over.[16]

Another great plant poison comes from the corn lily, which makes a chemical called cyclopamine. Cyclopamine doesn't hurt the plant at all, but it has a very strange effect on sheep. I'll give you a second to guess what that is. Your hint is that the name for cyclopamine comes from the one-eyed Cyclops of Greek mythology.

Got your guess? Okay.

Cyclopamine has no effect on an adult sheep, but if a pregnant sheep eats corn lily on the fourteenth day of fetal development (not before or after), the cyclopamine within the plant will block one particular set of genes from doing what it needs to do inside the sheep's developing fetus.[17] That's it. That's all this drug can do. One little step on one particular day in the fetal development of a sheep, but it's a whopper. The fourteenth day of fetal sheep development happens to be the day when cells in the fetus's head that will one day become eyeballs split into left and right halves. That step is controlled by the gene that is blocked by cyclopamine. What all this means is that if the mother eats corn lily on the fourteenth day of her pregnancy, then four and a half months later she will give birth to a one-eyed monster. As a result, any sheep flocks that try to stay in the corn lily's area will be unable to reproduce and, within a few short years, will die off from old age and leave the plants alone.*

Did you guess right?

* The disease in sheep is called cyclopea or, if you prefer bigger words, synopthalmia (Welch et al. 2009).

As wonderful as targeted drugs like cyclopamine are, they have the drawback that they might work on just a subset of the herbivores that are trying to eat the plant. Cyclopamine works on sheep, but it might not work on grasshoppers, for example. Many plants that use targeted poisons get around that problem by secreting a whole cocktail of them, with the hopes that something in there will hurt most herbivores somehow. Other plants use a more measured approach. They wait to see what kind of animal is eating them and then secrete the appropriate poison in response. It might seem incredible that a plant can do this, but it's true. The chewing bite of a caterpillar on a leaf of barrel clover will make the plant produce jasmonic acid, whereas the tiny piercing bite of a spider mite will cause the plant to make salicylic acid instead.[18] Plants might look like serene, inanimate objects, but they know how to get by in a world of gluttony.

Remarkably, plants can even communicate with one another about herbivores, so that a plant can start making poisons in anticipation when an herbivore attack has begun nearby. When a sagebrush plant from the southwestern USA, for example, gets chomped on by a herbivore, it releases some chemicals into the air. When those chemicals waft over to a nearby plant, that nearby plant responds by producing antiherbivore toxins in its own leaves so that it will be ready when the herbivores get there.[19] Why a plant should send that signal out is a bit of a mystery, since a selfish plant has no reason to help out its neighbors, but there are a few possible explanations. First, the plant may be secreting those chemicals to hurt the herbivores, and then nearby plants may just catch a whiff and react. Or maybe that first plant releases those chemicals as a way of quickly sending a message to all its other branches through the air. In other words, maybe it's communicating with itself, and then other plants are merely eavesdropping on the conversation. These are good guesses, but recently researchers showed that plants react more strongly to the smell of their clipped brothers and sisters than they do to more distantly related plants of their own species. That suggests that plants really may be sending information on purpose, trying to help out closely related individuals that have many of the same DNA strands they do. Future research will tell how it all fits together, but it's clear that plants have many more tricks up their sleeves to stop animals from eating them than most of us would ever have imagined.[20]

As a twist on the theme of plants talking to one another, the tobacco plant may have my favorite antiherbivore strategy. When it gets fed upon by the tobacco budworm caterpillar, it releases chemicals into the air. But in this case, the tobacco plant isn't crying out to another plant. It's sending out a signal for help, almost like Commissioner Gordon sending out the bat signal on a rough night in Gotham. The chemicals waft out into the air, and in response, parasitoid red-tailed wasps come swooping in to the rescue. They quickly lay eggs inside the caterpillars, dooming them to the torture of being eaten alive from the inside out by wasp larvae.[21] The wasps and plants gang up on the caterpillars, and Gotham is saved . . . until next time.

It's a constant battle, back and forth, between the plants that make sugars and the animals that want those sugars. For the most part, plants tend to get the upper hand, but there's one herbivorous animal out there that has really outdone itself. It's an inch-long animal that lives in the warm waters of the Atlantic Ocean. It resembles a snail with no shell, and it's called an emerald sea slug.

There are many kinds of sea slugs, and they're honestly some of the most breathtakingly beautiful, colorful things you'll ever see, but the emerald sea slug is even more fascinating for its diet than it is beautiful. It eats photosynthetic algae (sugar-producing organisms that are kind of like plants, but single-celled). Here's where things get crazy, though. When the emerald sea slug eats algae, it doesn't completely break the algae down to get the sugars out of them. Instead, it steals the molecular machinery from within the algae and moves that machinery close to the surface of its own transparent skin. Because the algae are green, that makes the emerald sea slug turn green. Then, astoundingly, the machinery continues cranking away, converting sunlight into sugars, inside the sea slug. The sea slug still has to eat algae from time to time, since the machinery wears down after a few months of use, but the result is an animal that performs photosynthesis—the near-impossible trick that Mr. Jani claims to have mastered.[22]

Even more incredibly, that solar-powered machinery isn't the only thing the emerald sea slug has stolen from the algae. Somehow it has also managed to copy a chunk of DNA out of the alga's genome and paste it into its own genome. That sequence of DNA is responsible for building some of the molecular machinery needed for photosynthesis. The sea slug still has to get most of the machinery by eating algae, but it makes a few bits of the machinery itself.*

The relationship between the emerald sea slug and the algae it feeds on adds a whole new layer of complexity to the story of DNA and meat robots. Suddenly, DNA sequences are jumping

* Honestly, the emerald sea slug's feeding strategy has got to be one of the best tricks ever accomplished by an animal. (But wait until the chapter on wrath, when you see what its cousin the blue dragon sea slug can do!)

between meat robots.* It's a reminder that DNA strands are the real players in the game of life, and that meat robots are just the biomolecular machinery.

The emerald sea slug isn't the only animal that gets a little energy hit from sunlight. There's an aphid that is orange when it grows up in normal conditions but grows up to be green if it's cold outside. That green color occurs because under cold conditions, the aphid produces photosynthetic machinery. The blueprints for that machinery come from a DNA sequence that looks to have been stolen from plants.[23] However, neither the emerald sea slugs nor the green aphids make enough energy themselves to survive completely without food.

To the plants, animals are often annoying predators that just make life harder, but that's not always the case. Some of the most successful plants have found ways to make animals work for them. I've already mentioned the bull's horn acacia, which enslaves ants to work as security guards, but that's just the tip of the iceberg. It's far more common for plants to enslave animals for something far better . . . and that's to satisfy their sexual urges.

Yes, plants have sex. Plants have boy parts and girl parts, and they use those to make babies. Pine trees, for example, have sex by releasing their pollen (sperm, basically) into the wind from male pine cones, and hoping some of the pollen happens to land on the ovaries that lie between the cracks of another tree's female pine cones.† Plants with flowers have sex by getting animals, like

* The technical name for DNA going from species to species is "horizontal gene transfer," as opposed to "vertical gene transfer," which is what happens when an animal passes on its DNA to its offspring.

† I have a good botanist friend who delights in pointing to people's pollen-covered cars in the springtime and telling them that a tree has just pleasured itself all over their hoods.

bees, to visit those flowers in exchange for nectar. In the process, the animals accidentally pick up pollen from one flower and drop it off at the next. Bees have basically become flying penises, meat robots under the influence of plant DNA.

Sex is the only reason flowers exist (which is why I sometimes smirk when I see people sticking their noses right into a bouquet of flowers: those are plant sex organs, after all). Many insects besides bees pollinate flowers, as do several birds, most notably hummingbirds. But what a lot of people don't realize is that some bats feed on nectar and pollinate flowers too. Those bats get their food by sticking their faces down into flowers while they hover, just like hummingbirds do, getting covered with pollen in the process. In fact, there are quite a few flowers in the world that don't use birds or insects at all but rely solely on bats to help them have sex.

One particular species of bat from the cloud forests of Ecuador, called the tube-lipped nectar bat, gets all its food from flowers. Because it's become so dependent on the plants, it has fallen under the influence of the plant's DNA, and this has made the bat into one of the strangest animals in the world.

Extend one of your arms out in front of you. Now stick your tongue out, and imagine your tongue could reach all the way to the tips of your fingers. The tube-lipped nectar bat can extend its tongue *three times* that far—one and a half times the distance from its head to its toe.

The plant that these bats feed upon is called *Centropogon nigricans*. At first, these bats probably enjoyed getting nectar from flowers much like those in your garden, but once those bats started relying on *Centropogon nigricans* for food, the bats' bodies began changing in response to the selfish demands of the plants.

It turns out that the deeper a bat sticks its face into a flower, the more pollen it will transfer. Plants with slightly deeper flowers forced bats to stick their heads deeper inside than plants with shallow flowers, and that gave deep-flowered plants a reproductive advantage. Over time, flowers got deeper and deeper. At the same time, only bats with long noses and tongues could keep up, so bats changed, locked in step with the flowers. Over evolutionary time, the bats grew longer and longer snouts, and longer and longer tongues. Today, as a result of a plant's trying to pass on as much pollen as possible with each bat visit, the tube-lipped nectar bat has the longest tongue, as a percentage of body size, of any mammal.[24]

That bat's tongue is a great example of the way plants have bent the bodies and behaviors of animals through evolution, for no other reason than to serve the selfish needs of the plants themselves. Since plants can make the sugars animals need, plants have the power to make animals do pretty much anything they want them to. In essence, animals are slaves to their stomachs. They have to eat: gluttony is the only way to survive.

Besides gaining security or a way to have sex, plants get animals to do their bidding for one other task as well—helping plants move around—and this is where humans get as manipulated as any other creature out there.

The main problem with being a plant is that you can't walk, so if you try to reproduce by dropping seeds on the ground, they'll just end up trying to grow right next to you. That's bad for the plant, since it puts it in direct competition with its own offspring for things like water and light and nutrients.

So instead of just dropping its seeds on the ground, many plants build a succulent, sugar-filled ball of deliciousness and hide their seeds in the middle of that. Animals come to take the food and end up carrying the seeds away too. That's what fruits are, and although it might seem like nature created avocados and apples and oranges and bananas to make humans happy, those plants really just wanted to get their seeds dispersed. Feeding animals such as humans was a convenient way to get that done.

In some cases an animal will carry a fruit away, eat it somewhere else, and drop seeds in the process. In other cases it will eat the seeds along with the fruit and then poop them out later, somewhere far from the host tree. In fact, many seeds germinate better if they've been through the digestive tract of an animal than if they haven't.[25]

There are thousands of different kinds of animals used by plants to disperse seeds. Many of them are things you'd expect, like toucans, parrots, monkeys, and fruit bats. But there are also more than two hundred different kinds of fish that eat fruits. That might sound impossible, but when you consider that the yearly flooding in the Amazon rainforest can raise the water level more than sixty-five feet, it's not so hard to picture a fish swimming up to a branch and biting off a fruit. Some fish can carry seeds more than three miles before pooping them out. To the plant, that's just as good as having a flying animal do the job.*

For animals to do that work for the plants, the fruits need to be as attractive as possible, and that is why fruits are so damned delicious. No hydrogen cyanide, no one-eyed mutant babies, no

* The fish, called *Colossoma macropomum,* looks just like its cousin the piranha, but instead of razor-sharp teeth, it has rounded ones for eating fruit (Anderson et al. 2011).

paralysis. Just sweet fruity goodness. Papaya, watermelon, mango, cherries . . . the list goes on. In some cases we humans have just taken the fruits pretty much as they are from nature (mangoes, for example), but in other cases we've used selective breeding to make the fruits even more delicious than nature made them.

This is true of apples and oranges and many other fruits. Wild bananas, for example, have numerous large, stonelike seeds within them. Humans, over the millennia, have cultivated some kinds of bananas so that the seeds have been reduced to nonfunctional black specs. That's taken away the banana plant's ability to pass on seeds through its fruits, but since humans grow banana plants all over the world just to eat the fruits, you could argue that their fruits are still doing their job of ensuring that banana plants survive and reproduce.[26]

Plants make their own food, but gluttony's still been an important factor in their evolution because it's allowed them to extend their influence to the animal meat robots around them. But the importance of gluttony to the natural world is even more striking when the gluttons eat other gluttons.

Some of nature's most charismatic animals are predators. You can tell because predators make for cool tattoos: polar bears, Bengal tigers, great white sharks, tarantulas, barn owls, saltwater crocodiles, rattlesnakes, orcas, wolves, giant squid, praying mantises . . . There are literally thousands of kinds of predators, living in all kinds of habitats, killing other animals every single day in the name of gluttony.

From an energetic perspective, eating meat isn't really very different from eating a plant: the molecular machinery of one living thing gets broken apart, then incorporated into the molecular machinery of another living thing. But as humans we see a very important difference between carnivores and herbivores. So far as we know, plants don't experience pain or fear, but when you kill and eat another animal, your lifestyle imposes suffering on other living things.

That distinction means a lot to us. We empathize with animals. Meat-processing companies spend millions to reduce the stress experienced by the animals they slaughter, and many people choose not to eat meat at all, just to limit animal suffering. But the overwhelming evidence from nature shows that other animals don't care very much about that kind of thing (at least when it's not happening to them). It's impossible to know, but I have a hard time imagining that a spider eating a fly gives it any more thought than a sheep gives the blades of grass it chews on.

Carnivores have to eat meat, so they kill animals all the time, but there's something strangely creepy about the fact that herbivores do this from time to time as well. For example, orangutans mostly eat fruit, but back in the 1980s, people started noticing orangutans killing and eating small (and heartbreakingly cute) primates called slow lorises.[27] In one well-described case, the orangutan smacked a slow loris out of a tree, crawled down to pick it up, bit into its skull, sucked out the brain and eyeballs, then ate the palms of the hands, the genitals, and finally the internal organs and skin. For some reason, that's creepy to me. An eagle does that and I don't blink an eye, but when an orangutan shows its dark side, I get all freaked out. The fact is, though, meat's a great source of food. I don't know why I should hold orangutans to a different standard than anyone else. Just because orangutans are usually vegetarian doesn't mean they've chosen that lifestyle to avoid hurting animals.

Eating meat has a lot of advantages over eating plants. For one thing, muscle tissue is packed with energy, so you get a bigger bang for your buck than you do eating leaves. Also, whereas plant-eating animals often need to eat many different kinds of plants to get all the nutrients they need (and avoid getting too much of any one plant-produced poison), a predator can find just about everything it needs in just about any animal. A mountain lion doesn't need to eat deer one time and then rabbit the next. It just needs a certain amount of meat. That allows it to prey on whatever animals happen to be abundant, and that gives it flexibility in an unpredictable world.

If we're going to talk about gluttony in predators, an obvious starting point is the wolverine, since its scientific name, *Gulo*, literally means "glutton" in Latin. A wolverine is badass. It weighs somewhere between twenty and forty pounds but can take down an eight-hundred-pound moose by jumping on its back and severing the tendons in its neck. You can think of wolverines as ninja mini-bears with semiretractable claws. (Seriously, that's not far off.) In one instance, a single wolverine was recorded to have killed ten reindeer in one day. (Sometimes wolverines stash the food they don't eat so they can come back to it, but that must surely have been overkill.)

As great as they are at hunting, wolverines are even better at finding carcasses of animals that have already died and eating those. In fact, wolverines have been known to track predators like wolves and lynx, wait until those predators make their kill, and then steal the carcass for themselves.

Using a combination of killing and scavenging, wolverines do very well for themselves. An impressive number of different animal species have been found in the stomachs of wolverines:

moose, elk, caribou, deer, foxes, lynx, hares, marmots, ground squirrels, porcupines, beavers, voles, lemmings, shrews, magpies, hawks, ptarmigans, fish . . . even seals, walrus, and whales.[28] That's the flexibility you get by eating meat. Wolverines can eat whatever they get their claws on, dead or alive. But for all the glorious gluttony of *Gulo*, it's far from being the most gluttonous predator out there. That title goes to an animal that at first might seem much less dangerous, the shrew.

Shrews make their living eating insects, worms, and the carcasses of larger dead animals. They're small, but don't underestimate them. Shrews are arguably the biggest gluttons in the animal kingdom. Sure, a 3,750-pound elephant eats far more food than a 0.1-ounce shrew does, but when you take body size into account, the tables turn. The roughly 225 pounds eaten each day by that elephant make up around 6 percent of its body weight. Compare that to a recorded maximum 384 percent of body weight consumed by a shrew in one day and there's no contest.[29]

Shrews and elephants are part of a more general trend: as you look across mammals of different sizes, the smaller they are, the more energy they use *per unit mass*. Since shrews are the smallest mammals, they're the most gluttonous. Incidentally, if you gathered enough shrews to equal the weight of an elephant, those half a million individuals would collectively eat sixty-four times more food than the elephant would. Pound for pound, nothing's so gluttonous as a shrew.[30]

But there's another way to measure gluttony in a carnivore, and that's to ask what animal *kills* the most other animals to survive. Something like a wolverine might only make one kill a week, and the shrew might get by on just a few earthworms a day. Who does the most killing in the name of gluttony? For this one, we go to the other end of the size spectrum, to the largest animal in the world.

Blue whales aren't just the biggest animals on Earth, they're the biggest animals to have *ever lived* on Earth. They're bigger than dinosaurs. A single animal can weigh more than 360,000 pounds, yet it feeds on shrimplike creatures called krill that weigh less than an ounce.[31] As a result of that size mismatch, blue whales have to kill a lot of prey animals to survive. Fortunately for the whale, though, krill routinely swim around in dense clouds of individuals. That makes getting enough krill as easy as swimming to the right place and opening your mouth. To get that krill without drinking too much seawater, though, the whale does something analogous to the way you separate your spaghetti noodles from the water you cooked them in. The whale swims into a cloud of krill, closes its mouth around that cloud, then pushes water out of its closed mouth with its tongue, forcing the water through hairlike baleen it has around the seal of the mouth (where you'd find teeth in most mammals). With the water expelled from the mouth through the strainer-like baleen, the whale swallows the krill, then sets out preparing for the next mouthful.

On average, that strategy lets a blue whale consume roughly 2,500 pounds of krill per day.[32] That's less than 1 percent of its body weight (consistent with the general trend that larger animals consume less food per unit weight than small animals do), but it means more than 500,000 lives are snuffed out each day by a single animal. That's a lot of death, especially since blue whales are generally portrayed as gentle giants. I suppose, though, that most people don't worry so much about the suffering of krill the way they do about the slow lorises that orangutans smack around. That's what makes these kinds of comparisons tricky. How many krill is a slow loris worth? Are intelligent animals more valuable than stupid ones, or does it just depend how cute they appear to us? Is it body size? Is it something else?

I don't mean those as rhetorical questions. In fact, those kinds of questions led me to be a vegetarian for several years after undergrad. I avoided meat because I felt that it would be hypocritical to say that some animals, like bats or dolphins, for example, should never be harmed, but then eat cattle and salmon as if they grew on trees. Even though the idea of killing some species *felt* worse to me than the idea of killing other species, that didn't seem like a good enough reason for me to start separating animals into edible and untouchable groups. I understood that the whole spectrum of animals, from slow lorises to krill, know to move away from things that can hurt them. To me, that seemed to imply that all those animals had some knowledge of pain, suffering, and fear. It just didn't seem consistent for me to inflict those experiences on some animals but not others. To me, giving up meat altogether seemed like a way to live less hypocritically.

There is a way to eat meat without inflicting suffering on other animals, but that was not a solution I was at all interested in—eating things that died naturally. To humans, nothing's more repulsive than chowing down on a bloated corpse that's been ripening for a couple of days. There are plenty of living things out there, though, that enjoy nothing more, and those gluttons are vitally important parts of the living world.

Let's say a person dies suddenly from a heart attack while they're on a walk in the woods, and no one finds them for a few days. The instant that person's body shuts down, the calories inside it are no longer defended. So, less than *four minutes* after death, creatures start taking that person apart.[33] Whole colonies of bacteria that had been living in the gut, helping the human digest their food, suddenly find that the walls of the digestive tract are no longer protected by an immune system. They immediately go to work on the human itself, feeding and reproducing at a feverish pace. The walls of the digestive tract break down, leaking bacteria into other parts of the body, to begin preying on the other internal organs. As they feed, those bacteria release methane and sulfur-rich gases in the process, causing that characteristic dead-mammal smell to emerge from the body. It's a smell so disgusting to humans it can make them vomit, but that same smell draws in other animals, like foxes, crows, flies, and beetles, that can make a nice little meal from a rotting human corpse. Just how long it takes for a human to be reduced to bones depends on a lot of factors, most notably temperature, but at around 68 degrees Fahrenheit, sixty-five days is a good ballpark estimate.* It takes even less time, though, if the body is in water, if large scavengers have taken the body apart, or if the person died from a large wound that drained blood and allowed microbes easy access to the body.

Other animals, in other places, decompose more quickly. One of my favorite bat caves in the world is Bracken Cave, near San Antonio, Texas. It's a particularly hot cave, and there are *millions* of bats in there. Beetles thrive on the floor of the cave, feeding on bat droppings most of the time, but should a baby bat fall to the cave floor before it can fly, its tiny body will be reduced to bones in less than ten minutes. Walking through that cave, you can see that guano (bat droppings) covers the cave floor, but if you look

* Vass (2001) gives a neat equation to determine the number of days it will take a body to skeletonize: just divide the number 1285 by the temperature in Celsius. At 68°F (20°C), that gives 64.25 days.

down carefully at it, you'll see what look like white pine needles everywhere. But there are no pine trees outside Bracken Cave. Those are the wing bones of bats that have been broken down by scavengers. Bracken Cave is one of the best places to see decomposers at work.

In nature, those decomposers are the last in the chain of energy traders. If you could follow a single calorie of energy out of a cave beetle, back through time, you'd get a convoluted but unbroken chain: the beetle stole energy from the baby bat, who got her energy from her mother's milk, who got her energy by eating a moth, who as a caterpillar got his energy by eating a corn plant, which got its energy from the sun. If you continued playing the sequence out, like a movie going backward in time, it would end with a ray of light going back on its eight-minute-and-eighteen-second trip between the Earth and the sun. That's a movie I'd love to see.

In fact, you could do the same with any calorie inside any animal, and the movie would always play through to the same beginning.[34] You could watch the energy bounce among organisms, through parasites, predators, and prey, but it would always ultimately have come from the sun. If you could play all those movies backward together at the same time, they'd all converge into a broad beam of light. And if you stopped that movie suddenly, then started playing time in the right direction, you'd suddenly see the flow of energy through the living world as it exists around you right now. Every day, rays of light shower down on our planet. A few of them get harnessed by plants and enter the food web. From there they really might go anywhere. It's all totally predictable, since we know in what direction the energy will flow, but where any particular sunbeam's energy will end up is anyone's guess.

\\\\\\

My vegetarianism lasted for three years or so, but around the time I went on that botfly trip to Belize, I started eating meat again. It wasn't the botfly per se, but that was part of it. My relationship with the living world was changing as I learned more about it, and it was starting to dawn on me that I was still inflicting pain on animals. I'd killed the botfly, after all, but it was much more than that. Whenever I drove my car in the summer, insects got splattered on the windshield. Whenever I ate produce grown on farms, I was reaping the benefits of habitat destruction for the animals that had once lived where the farm now stood. Even a lot of the science I was learning came on the backs of animals, harmed or even sacrificed in the experiments that I was reading about in scientific journals. Sure, I wasn't eating them directly, but my lifestyle was inflicting pain and suffering on animals. If I was going to live up to what being vegetarian was supposed to be about, I had a long way to go. I was going to have to make sweeping changes to my life well beyond what I ate, and frankly it just seemed like a suite of lifestyle changes I simply didn't want to make.

I realized that my vegetarianism hadn't made me any less hypocritical than I'd been before, so I just let it fall away. I went back to eating meat, though much less than before, and with a very different feeling about carnivory than I'd had before. The experience wasn't a waste of time. It was just part of my education.

As I spent more and more time studying nature, chasing bats through tropical rainforests in Belize and Costa Rica, I started thinking of the food I ate as my connection to their world. Like a bat, I am built to get calories by taking them from other living things. I don't get angry at the frog-eating bat for cutting short the lives of frogs, so it was strange to lament my own need for calories. The idea of sparing animals out of sympathy is such a human idea; that doesn't make it a bad idea at all, but it's important to realize it's not an idea that exists in the nonhuman natural world.

Part of the popular image of Prahlad Jani that has emerged from the YouTube interviews and newspaper articles about him is that he's connected to the world so deeply that he doesn't need to eat. That's ironic. Eating is probably the most concrete way in which we participate in nature.

A few thousand years ago, Shelby and I would have needed to find food in the environment around us to feed ourselves and Sam. Now it's as simple as picking up some chicken breasts on the way home from work, but we're still part of that flow of energy, from sun to decomposer. The landscape is changing, though. A lot of people are scared about factory farming and genetically modified foods, but I'm still optimistic about Sam's nutritional future. In the last few thousand years, humans have domesticated more than two hundred different food crops, from agave to watermelon, making some more nutritious, some more resistant to disease, and some better able to grow in the high densities that you'll typically see on farms.[35]

Like the acacia plant that has a better life thanks to the security guard ants it domesticated, we thrive because we have manipulated parts of nature too. Humans have been playing by the same rules as the acacia plant for centuries, and it's the only reason we've succeeded like we have. Until recently, those changes came through selective breeding, and now that process also happens through genetic engineering. That doesn't scare me.

Had Sam been born a hundred years earlier, his life expectancy would have been fifty years, but because he was born in 2011, his life expectancy is closer to eighty, and I wouldn't be surprised if he made it to one hundred.[36] Sam's access to nutritious food is probably the greatest reason he should expect to live longer than any of his ancestors ever did. (That, plus protection from the kinds of parasites I talked about in the chapter on sloth.)

There's another reason we need domesticated crops and livestock, though. By the year 2050, Sam's going to have to share the Earth with 9 billion other people.[37] Without advanced techniques for food production and distribution, that would be impossible. You can't feed 9 billion people if you expect all of them to go out in the woods looking for edible shrubs. Like Mr. Jani, we all have to eat. Developing crops is the only way for Sam, or any of the other 9 billion of us, to do that.

5

ENVY

Of Thieves and Sneaker Males

After two days of labor, Sam still wasn't born. Shelby was working as hard as anyone could, but it was as though something was physically holding him back. When Shelby pushed, Sam would descend far enough that you could see the top of his head but then immediately bungee right back up. After two hours of pushing, Sam still wasn't progressing, so our doctor and doula recommended we think about a C-section.*

The choice to have a C-section was a heartbreaking one for Shelby, since she'd always dreamed of having her babies the

* A doula is a nonmedical person who assists before, during, and after labor and childbirth, acting as an experienced resource for the family. To me, the biggest benefit of having a doula was just having someone there the whole time saying, "This is normal."

old-fashioned way. But Shelby's instincts as a mom kicked in well before Sam was born, and once she could see that Sam needed help, she had no hesitation about the choice to have surgery.

In the room where the surgery happened, there was a curtain across Shelby's neck, so she wouldn't be able to see her belly being sliced open. As surgery began, I sat with Shelby on the "head" side of the curtain with an anesthesiologist and our doula, while other doctors went to work outside my view. (Ironically, there was some art on the ceiling of the room which, for hygiene reasons I assume, was laminated. That plastic surface turned the whole art piece into a giant mirror, meaning Shelby, of all people, could watch as much of the surgery as she wanted to.)

It all happened in a matter of minutes, and once Sam was out, the medical team put him on a blanketed table and invited me to come meet him. I stood up and walked over, bracing for what would surely be one of the most important moments in my life.

At first I wasn't seeing Sam so much as the parts of him. His legs were crooked and bony. His head was cone-shaped, bent and compressed from the hours it had spent wedged in Shelby's pelvis. I looked at his chin and could immediately see my great-aunt Claire's mouth on him.* It was clear to me that he was mine, and yet he was still so unfamiliar. I couldn't comprehend that this person could be my son. The whole thing overtook me, like falling into water and trying to get oriented in the bubbles so you can start swimming upward.

Sam was kind of squirming, with his face all scrunched up.

* Claire Riskin was my father's father's sister. Roughly one-sixteenth of Sam's DNA is the same as hers.

I thought maybe he might be uncomfortable, being held and prodded by all those doctors around him, so I put my hand on his chest. Some friends had once told me that a baby can learn its father's voice while it's still inside the womb, and can recognize that voice immediately after birth. So I tried speaking to him.

I spoke in my warmest, calmest voice: "It's okay, little guy."

Sam froze immediately, opened his eyes, and listened. That was our first contact. It was wonderful . . . but it only lasted a moment.

One of the doctors gently asked me to remove my hand, and as I did, I realized that the team of doctors and nurses swarming around the room hadn't slowed at all. I looked back at Shelby and saw her open abdomen on the operating table and the placenta sitting in a metal bowl next to her. Then I looked back down at Sam and realized he was purple! I hadn't even noticed it, but Sam still wasn't breathing.

The doctors kept working on him, inserting a tube down his throat to clear mucus, but the suction machine wasn't operating properly, and the intercom system one of the doctors was using to request a backup tube wasn't functioning either. Our own doctor calmly covered Sam's mouth and nose with a mask and pumped air into it with a handheld bulb. With each sequence of pumps, his skin turned less blue for a moment, but then she'd back away to let him take a breath on his own, and he would slowly go purple again. No one looked panicked—not even the doctor struggling with the intercom system—but everyone looked very, very serious. I couldn't tell from their expressions whether this was all normal or if I should be terrified (the doula was out of my sight), so I just stood there and watched. I was totally helpless. I really didn't know if my son was going to die. Seconds ago I had been full of joy, and now suddenly fatherhood was totally terrifying.

It was like that for several minutes, until Sam finally drew in that first breath. When he did, the exhalation that came from all the doctors around him showed me just how scared they too had been. As for me, I could barely stand up. Sam's first breath came *six minutes* after he was born. I didn't know what to feel. I was still disoriented, wanting to swim upward but not yet knowing which way was up.

Apparently, once a baby gets that first breath out of the way, it usually keeps breathing, so once Sam's lungs had started up, my doctor smiled, put him in my hands, and told me to take him over to meet his mom. I pulled him against my chest, supported his head with two fingers, and carried him back to Shelby. A minute ago I didn't know if Sam would die and now everyone was acting like everything was normal. I sat down, smiled with Shelby at Sam's scrunched-up face, and then sobbed uncontrollably, like I can't remember ever crying before.

Our doctor told us afterward that Sam's delay taking that first breath was fairly normal. Babies often do that if they're born via C-section, and research hasn't shown any long-term consequences from that kind of ordeal. Because she'd been pumping air into his lungs, he'd gotten plenty of oxygen that whole time. Our doctor also told us that when Sam was removed from Shelby's uterus, they'd seen that the umbilical cord was wrapped twice around his neck. It hadn't strangled him or anything, but she supposed that could explain why Sam didn't progress further during labor. I have no way of knowing whether Sam would have survived birth without medical interventions, but the whole experience was about as intense as anything I've ever been though in my life. On the one hand, I was overwhelmingly grateful that he was alive, but the helplessness I'd felt before that first breath left me feeling that I might not have the skills I'd need to raise this kid. It was exciting, but the stakes seemed unimaginably high. Fatherhood had turned every single one of my emotional dials to eleven.

Looking back, though, I can't help but see the way I experienced that day as textbook animal behavior. Here's the father meat robot, following orders from his DNA molecules, releasing stress hormones with all the predictable physiological effects, in response to seeing his offspring in danger.

And I still see myself like that to this day. Even though it feels like magic when Sam does something new, or even when he looks me in the eyes and smiles, it's all just biology. I know that no matter what it felt like in those first six minutes, the emotions I felt the day he was born were really no more special than the hormones that guide a bird through building her nest.

\\\\\

Envy is a big part of having a new baby, because you're constantly comparing notes with other new parents. For Shelby, it was most pronounced when she spoke with other mothers who didn't have C-sections. In the first few months after Sam was born we talked about it often. It didn't matter what our doctor had said about the umbilical cord. Shelby couldn't seem to shake the gut feeling that if she'd just been tougher or more patient, somehow her C-section could have been avoided. For so many of the mothers we'd talked to, it had just worked out, so Shelby's instincts kept telling her that she'd done something wrong. With a couple of years between her and the experience now, she's not so sensitive about it as she once was, but I know that it still bugs her.

Now the comparisons we make with other families tend to be less about childbirth and more focused on child development. And we know that other parents are making comparisons too. Sam woke up four times again last night, but our friend's kid, Julia, has been sleeping through the night for two months already. Sam's learned to count to three, but Olivia can count backward from ten. Shelby and I know that the rate of development is different for different kids (and we've read several books and scientific papers about development—we're scientists after all), but it's still really hard not to have an emotional reaction sometimes when you hear something about another person's kid. When Sam's ahead, we feel smug, and when Sam's behind, we speculate that it's only because other aspects of his development are so advanced. But I'm embarrassed that we make those comparisons in the first place.

It's so stupid, really. Shelby and I know people whose kids have had cancer, for God's sake. We have friends who have miscarried. You would think that we could just count our blessings and move on. Sam is a healthy, happy kid and he's doing great. But there's something about the experience of raising a kid that makes you listen very carefully to see what's going on with all the kids around you.

I suppose that from an evolutionary perspective, paying attention to the parenting experiences of other people is a good way to see how well your child is doing. After all, as we discussed in the chapter on greed, your kid is going to have to compete against other kids down the road, so how he compares to those kids is going to matter someday. But worrying about how many words Sam knows at eighteen months compared to the Joneses' kid is a waste of energy. I really don't like that I experience envy at all. Envy has been described as "feeling negatively about someone else's success," and I don't want that to be part of how I raise Sam.

Envy is part of the human experience everywhere in the world, and it can motivate all kinds of crimes, from vandalism to murder. But there's really no better symptom of envy than theft. When humans experience the desire to have something that another person has, they often just take that object of desire away. Envy in animals is probably much less self-reflective, but theft among animals is everywhere.

It might seem impossible that animals can steal from one another, since animals don't have money or property law, but tell that to a cheetah who does all the work to sneak up on, chase, catch, and kill a gazelle, only to have that carcass taken away by hyenas. Since the cheetah has invested time and energy in that carcass, biologists consider it fair to call it theft when it's taken away.*

Some animals are more vulnerable to theft than others. Animals that swallow their food immediately, like snakes and frogs, don't have much of a problem with other creatures intercepting their calories before they're eaten. But lots of animals need time to ingest their food after it's captured. Cheetahs are great examples of this, and so are ants.

Leptothorax ants live in colonies of a few hundred individuals,

* Theft is really a type of parasitism, since (as you'll remember from two chapters ago) parasitism describes any relationship between two organisms in which one gets a benefit (the parasite) and the other pays a cost (the host). In the case of parasitism by theft, biologists use the word *kleptoparasitism,* and talk about animals that are kleptoparasites of other animals. You might recognize the prefix *klepto-* from the word *kleptomania,* the disorder that causes people to steal pathologically.

and for those colonies to function, they need a food transportation system. Food is swallowed by workers, which return to the nest, then regurgitate that food to feed the growing larvae. Some of the larvae grow up to become workers themselves, while others grow up to be soldiers that defend the colony. At the center of it all is a single queen—the only one who can reproduce, and the mother of all new ants born into the colony. She's the only hope any ant in the colony has of getting its DNA passed on to future generations, so every meat robot in the colony ultimately works for her survival. Because a *Leptothorax* ant colony is large and complex, it takes time for the colony to process food, and that makes it possible for thieves to work the colony over.

One such thief is the queen of a different species of ant, called *Epimyrma* (the Latin name means "above ant"). She sneaks silently into the *Leptothorax* nest like a cat burglar. When the first *Leptothorax* guards attack her, she outmaneuvers them like Jason Bourne, then stings them with just enough venom to stun them. She doesn't kill them because they'll be useful to her later. Next, she rubs secretions from *Leptothorax* ants all over her body, to mask any trace of her foreign scent. Once that's done, she can move through the colony undetected. Because she smells like a *Leptothorax* ant, the members of the colony have no way of identifying her as an intruder. Cloaked now, the invading female finds her way to the *Leptothorax* queen's chamber, approaches the undefended matriarch, and throttles her slowly with her mandibles, until she is dead.[1]

With the old queen dead, the *Epimyrma* queen takes her place as the new leader of the *Leptothorax* colony. She lays her own eggs, then lets the workers of the colony care for her and feed her offspring, unaware that they've been enslaved. In doing this, she's not just stealing food from the *Leptothorax* colony. She's stolen the colony itself.

This kind of theft is practiced by at least two hundred different kinds of ants. However, not all ants do their stealing the same way. For instance, the ant *Polyergus* will mount "slave raids" in which roughly 1,500 individuals move together to attack the colony of an ant called *Formica*.[2] The marauding ants will capture as many *Formica* larvae as they can find, then bring them back to their own colony. When the *Formica* larvae emerge as adults within the *Polyergus* colony, they just start working, with the mistaken impression that they are where they belong. The kicker is that *Polyergus* ants have become so dependent on their slaves that they can't live without them. *Polyergus* colonies always have *Formica* ants in them. Without slaves to do the work, *Polyergus* ants can't survive. It's just another example of an entire species whose survival depends on causing hardships for members of another.

Unlike ants, spiders are usually solitary, but their habit of storing food before eating it makes them vulnerable to theft too. An orb-weaving spider, for example, processes food relatively slowly. The spider spins a web and sits at its center, waiting for the vibrations that happen when an insect gets tangled up. If it senses those vibrations, the spider heads over to the struggling insect, wraps it in silk, stabs it with venom, and leaves it hanging, in order to continue hunting. As the spider heads back to its perch at the center of the web, the insect lies helplessly bound up, spider venom chemically digesting its internal organs. Later, when the spider wants to get the calories out, it can go back to the now-dead insect and slurp out the liquefied insides like a smoothie.

The spider *Argyrodes* could do all that work for food itself too, but it chooses not to. Instead, it hangs out at the periphery of an orb-weaving spider's web. Orb weavers are impressively large, around two inches long, while *Argyrodes* spiders are only a quarter to half that size. Sometimes, when a small insect gets tangled in the orb weaver's web, the *Argyrodes* spider runs over and steals it before the orb weaver gets to it. That's usually not a big problem for the orb weaver, though, since it most often happens with insects that are so small that the orb weaver probably would have just ignored them anyway.

But sometimes *Argyrodes* goes for larger prey. When a nice big, juicy insect hits the web, the orb weaver immobilizes and injects it with venom as usual. The *Argyrodes* spider can observe the capture by feeling the vibrations in the web. Then it waits for the large spider to get distracted by a subsequent prey item. Once the orb weaver heads off to wrap up that second prey item, the *Argyrodes* spider scurries in, cuts the original meal out of the web, and runs away with it.[3]

To add insult to injury, *Argyrodes* spiders don't always eat the food they steal. In some cases, males have even been seen giving stolen food away to females as gifts, to avoid getting eaten during sex.[4] We discussed in the chapter on lust how the preferences of females often create life-threatening hurdles for males to jump over before the females will agree to mate with them. A demand by females for gifts that have to be stolen from a predator big enough to eat the male is a perfect example.

The strategy *Argyrodes* spiders use to steal from other spiders relies on never getting too close to them, but the spider *Curimagua* has no reservations about walking right up to a spider thirty times its size and stealing food right out of its mouth. It's only about a millimeter long, but does its thieving from a one-and-a-half-inch-long spider called *Diplura*. Those larger spiders make funnel-webs, and when a large prey animal, such as a grasshopper, beetle, or even frog, walks near the opening of the funnel-web, the *Diplura* launches out, strikes it with its fangs, then drags its prey back to the opening of the funnel to suck its insides out.

As the big *Diplura* spider starts feeding, the *Curimagua* spider walks right over to its mouth and starts eating alongside it.[5] In fact, the tiny thief can apparently walk anywhere it wants to—in the funnel-web, or even across the *Diplura* spider's eyes. The larger spider doesn't seem to pay any attention to it all, perhaps because it's so small that the calories stolen cost *Diplura* less than the calories it would have to spend catching and killing the intruder.

It's a strategy that works for the small spider. So well, in fact, that it has lost the ability to live anywhere else. Unlike other thief spiders, such as *Argyrodes,* that can make webs and catch their own prey whenever they need to, the *Curimagua* spider can no longer hunt for itself. So far as I know, it is the only spider species in the world, out of more than 44,000, in which individuals don't do their own hunting.[6]

Another unique spider that commits acts of thievery is a very special jumping spider called *Bagheera kiplingi*. (That Latin name will stick in your brain if you've ever read Rudyard Kipling's *The Jungle Book* and you remember Bagheera the black panther.) *Bagheera kiplingi* is the only known vegetarian spider in the world. Remember those acacia plants that feed nectar to ants in exchange for defense? Well, *Bagheera* lives on those same acacia plants, eating the nectar that the plant has made for the ants. As you can imagine, this doesn't thrill the ants, so the spider has to constantly hide from them, jump out of the way when attacked, or else hang from a thread so no ant can reach it. The spider isn't a *strict* vegetarian—it dines on larval ants from time to time, so it's also technically a predator of the ants—but *Bagheera* gets the vast majority of its food by stealing nectar from ants, rather than by killing them.[7]

It's not just creepy crawlies that deal with theft. It's a problem for big animals too. Many African predators, for example, supplement the food they kill themselves with food they steal from other predators. Hyenas are famous for stealing from lions, but things aren't as one-sided as *The Lion King* might have led you to believe. Lions are aggressive carcass thieves too, and it's quite often the case that hyenas hunt and kill their own food only to have it stolen by a pride of lions.[8]

Theft is more than a nuisance for these animals: it threatens their survival. Kills don't happen every day, so predators sometimes go several days without food. That's why having a meal stolen can create big problems, especially right after an animal burns a whole bunch of calories running down and killing its prey. A cheetah, for example, will hide its food as quickly as possible after a kill, but once competing carnivores find it, the cheetah has to move on. This means that even with an abundance of prey to feed on, cheetahs may not be able to use certain habitats because of competition with other predators. In fact, if a cheetah so much as hears the calls of hyenas and lions, it will stop hunting and move to another location, presumably to save itself the ordeal of having food taken away.[9]

Similarly, African wild dogs are endangered, with fewer than six thousand individuals left, so nature reserves have been set aside to protect them and other endangered animals where they live. But because hyenas and lions do so well in those same reserves, the African wild dogs can't seem to get a break. One study in Zimbabwe showed that African wild dogs inside one such park had their food stolen roughly twice as often as they did outside the park. As a result, they chose to spend most of their time outside the very conservation area that had been set aside for them.[10]

There's an appealing concept that nature is naturally balanced and self-regulating, and that the key to making nature thrive is for humans to take themselves out of the picture and let things come to their natural order. But the truth is that except for humans, animals have no concept of order and simply do their best to thrive as individuals. If animals in an ecosystem are left alone for long enough, a balance emerges, but when that balance is perturbed, things often don't return to the same equilibrium. As a result, if we stopped managing wildlife in the savannahs of Africa, it's not necessarily true that cheetahs or wild dogs would be able to bounce back. It was humans that brought their numbers down, but even if humans vanished, the problem of theft by other animals, or perhaps something else altogether, might prevent cheetah or wild dog populations from ever recovering.

Theft by the meat eaters of the African savannahs has special significance to us because we evolved as a species among them. If we could travel far enough back in time, we'd see lions and hyenas stealing food from other predators, but they'd also be battling with a third species for those carcasses, and that third species would be us.

To early humans looking for meat, chasing some lions away from a wildebeest carcass would have been dangerous, but it would surely have been a much more appealing strategy than the alternative of hunting the wildebeest themselves. For early humans, that second option would entail chasing the animal on foot for eight hours or more in the searing midday sun.

The fact that humans can run down a large animal that way sounds impossible at first, but persistence hunting, as it's called, has been documented in traditional societies around the world, including the Kalahari people of southern Africa, the Hadza people of eastern Africa, the Tarahumara and Navajo people of northern Mexico and the American Southwest, and the Aborigines in Australia.[11] Humans are built for running. Even people who sit at desks all day but do a little exercise a few days a week can typically finish a 10-kilometer run without extraordinary effort. In fact, each year thousands upon thousands of people complete marathons of more than 26 miles, and some even go farther, completing ultramarathons of 100 kilometers (62.1 miles) or more. The idea that our ancestors could run for eight hours straight isn't so far-fetched. And although big animals are faster than humans over short distances, even they can't run as far as a fit human can. Over long enough distances, humans are some of the best runners on the planet.

When a human chases a big animal under the hot sun, the animal being chased eventually overheats. Humans can cool themselves while running by sweating constantly, but large four-legged animals cool themselves by panting, and since those animals take only one breath with each set of strides, they can't pant fast enough while running to cool themselves down. Without a chance to stop and pant in the shade, the larger animal eventually overheats, collapses from exhaustion, and is then easily killed by the humans who pursued it.

In the 1980s, researchers spent a year living among a group

of around fifty Hadza people, who still subsisted off the land in eastern Africa, not yet having been swept up in the agriculture and urbanization around them.[12] What the researchers found was that the tribe did use persistence hunting but that members of the tribe always paid close attention to their environment for any signs of a fresh kill by other predators—circling vultures, or the nighttime calls of hyenas or lions. If such a clue was detected, the Hadza men would immediately run in that direction. Leopards and hyenas would run away as soon as the people got there, but lions were stubborn enough that they often ended up becoming part of the human meal. Over that year, 20 percent of all the dead animals brought back to the village had been scavenged from other predators, including meat from elephant, zebra, warthog, giraffe, wildebeest, and impala. Humans living traditionally in other parts of Africa, including Cameroon and Uganda, have also been observed stealing meat from predators. Assuming that those twentieth-century people used hunting methods similar to those used in the same places millennia ago, theft is probably a big reason for our success as a species. All the evidence points to humans being thieves for almost as long as we've been able to walk on two legs.

Do animals *really* steal because they're envious? That's a tough question. It's not quite impossible to determine whether or not an animal truly experiences envy, but it's pretty close. To demonstrate conclusively that animals are envious, you first have to demonstrate that they comprehend that other animals have their own experience. Then you have to show that the animal can measure a difference between what they have and what a competitor has. And then, finally, you have to show that this imbalance makes the animal react negatively. That's a tall order. Even if you see animals doing the kinds of things humans do when they're envious, like stealing from one another, you can't really *know* it's envy unless you do a rigorous set of experiments.

But those experiments *have* been done with capuchin monkeys, and the results are pretty compelling.[13] Researchers trained some monkeys to put a small rock in a researcher's hands. As a reward, the monkey would either get a grape, which they *loved*, or they'd get a little cucumber, which they found tolerable but far less exciting than a grape. To bring envy into play, the researchers fed monkeys their grapes and cucumbers in side-by-side cages, so the monkeys could always see what their neighbor was getting.

So long as the reward a monkey got was the same as the reward their neighbor got, monkeys played along, happily taking grapes or cucumbers in exchange for their work. But if a monkey got a cucumber while it could see its neighbor getting a grape, the monkey got angry. Monkeys would sometimes throw the cucumber back in the face of the researcher or else just stop playing the game altogether. In essence, monkeys refused to do the same work if they saw someone else getting paid more for the same job.

Now, what I've described so far might be monkey envy, but there's another possible explanation. Perhaps the monkey getting the cucumber is reminded of how awesome grapes are when it sees the grape next door. Maybe it's not envy but just that it's hard to eat crap food when you think about delicious food. Maybe rather than envy, the monkey just feels disappointment.

Well, the researchers designed their experiment to rule out disappointment so they could be sure that what they were seeing really was envy. The researchers did that by making sure that in those scenarios where both monkeys were getting cucumbers, they could both see a bowl of grapes right in front of them. The fact that monkeys didn't get mad those times showed that the thing upsetting the monkeys wasn't being reminded of how awesome grapes are. The trigger was clearly the difference between what a monkey got and what its neighbor got. The careful design of those experiments nicely showed that capuchin monkeys experience envy.

Dogs show the same tendencies. In another study, researchers put pairs of dogs side by side and asked them to "give a paw." The dogs obeyed the order up to thirty times in a row when researchers gave both dogs treats as rewards, or even when the researchers gave neither dog a treat. But if one dog got treats while the other one did not, the dog getting nothing quickly stopped obeying commands. Again, researchers made sure it was envy rather than disappointment by leaving the treats out all the time, right where dogs could see them.[14]

I suspect that many other animals experience envy, but experiments just haven't been done for them. (Experiments have been done with chimpanzees, but the results are less clear.)[15] There may be many other kinds of envious animals in nature, but without those kinds of experiments, you can't ever know for sure. When you see a pigeon at the park stealing a piece of bread from another pigeon, you can't tell if it was envious of the other pigeon's property or if it just wanted some bread. The fact is, though, that nature's a tough place because animals steal from one another. Splitting hairs about their inner emotional state sort of misses the point.

Besides, envy can be a good thing. If a kid admires the possessions of the doctor next door and that motivates her to go to medical school, I don't think that's a problem. If a person envies the peace and calm of a colleague from work and decides to take up meditation and yoga to try to emulate those traits, great! Looking around to see what others have and then trying to get those things for ourselves is a basic part of how human society works. The fact that capuchins and dogs have envy might be the small cost that comes with all the advantages of being part of a social species.

But envy does bad things sometimes too. And while theft is an obvious symptom, another one is infidelity. Think about it: coveting thy neighbor's partner can sometimes lead to affairs, and that can cause serious emotional pain for everyone involved. Animals cheat even more than professional athletes do, and I have to think that envy plays a role for them at least some of the time.

Take species with harems, for example, where you have a single male who gets to mate with a whole bunch of females. This always means that a bunch of other males have to sit on the sidelines and mate with no one. How could they possibly *not* be envious? My favorite example of this is the sac-winged bat I mentioned in the chapter on sloth, the ones I saw roosting outside my first vampire bat cave in Costa Rica and whose males are known for flinging feces and urine on the females they want to mate with. (Yes, *those* sac-winged bats.) So help me, if there's envy surrounding sex in the natural world, that is a species that would have it.[16]

Male sac-winged bats have a lifestyle in which one male gets to be the partner to a harem of up to seven females. Those females live with one another, and with one another's pups, on the side of a tree, cave, or building. Solitary males, called satellite males, roost nearby, without any females of their own.

The sac-winged bat got its name from the pockets it has in its wings, just above its elbows. (Here's where things get disgusting.) Right before heading out for a long night of hunting insects, a male licks out the sacs in his wings to clean them, then leans down, fills his mouth with his own urine, and spits that urine into his wing sacs. Next, he leans back down and rests his throat against his penis and quivers until a white droplet comes out of his penis and sticks to the fur on his chin. Then he transfers that droplet with his chin to the urine-filled wing sac, before repeating the procedure to fill the other wing sac.

Mixed together, the urine, saliva, and genital secretions produce a musky bouquet of odor. The male then hovers in front of a female in his colony and shakes his wings at her, much like a person would shake a salt shaker at their food, thereby covering that female with a sample of his sexy cologne. This behavior is called salting.

Males are advertising how healthy they are with the way they smell, and females use scent to decide whether or not the male is sexy enough to mate with. That's why males have to clean their sacs out each day and make a new batch. Bacteria living in the sac constantly break down scent molecules, producing other, less attractive odors. To a female, those bacterial odors are a sign that the male isn't quite up to snuff.

Every once in a while, a satellite male will salt one of the females, giving her a sample of his wares (sometimes he'll even salt the harem male!), but the harem male responds by chasing him back to the periphery, and salting on him for good measure.

It's pretty clear that the satellite males are just waiting for an opportunity to take over the harem. When researchers experimentally remove a harem male from his roost, it's always a satellite male that takes his place, as opposed to a dominant male from another harem, for example. And it's not just any satellite male.

It's a satellite male that had previously been sitting right at the periphery of that particular harem. Since there can be several harems close to one another on a wall, I think that's suggestive that a satellite male is focused on stealing one particular male's harem. It looks for all the world like envy to me, but I know it's impossible to know for sure until controlled experiments are done.

The really interesting thing about the harem mating system in sac-winged bats is that everybody is cheating. In a frenzy of paternity analysis that would make Maury Povich envious, researchers learned some dirty sac-winged secrets. Yes, harem males fathered more of the pups than satellite males did, but a harem male was the father of only about 30 percent of the pups in his own harem. Thirty percent! The other 70 percent came from a mix of nearby satellite males and a few males with harems of their own, whom females had obviously gone out to visit from time to time.

Harems are common in nature, and that makes life pretty tough for the males out there who just can't compete with other males. But there's an alternate strategy available: if you can't beat 'em, become a sneaker male. Sneaker males occur in all kinds of different animals, and their strategy is to avoid confrontations with the stronger males by getting creative.

Great Plains toads mate in the puddles that form after a big storm in the midwestern Unites States. Males initiate this process by getting into the water and croaking out into the night, to say, "The water's fine, ladies. You should come join me!"[17]

One thing about frogs is that the male has no penis to place inside a female. Instead, he'll hold on to her in the water, hugging her from behind in a sort of frog cuddle.* While he does that,

* This frog mating cuddle is called *amplexus,* which is Latin for "embrace."

she will release her eggs into the water as he releases his sperm on top of them. That is how frogs have sex, but first they have to meet up.

A female Great Plains toad won't approach just any croaking male. She prefers a large male, and there's a biomechanical link between how big a toad is and the pitch of its voice: bigger toads have deeper voices. So by calling out with a deep, booming voice, a big toad can advertise his size. This sort of screws the little males over, since their tenor tones just don't impress the females, and there's no way to fake a deep voice. (No matter how well you play a trumpet, it will never sound like a tuba.)

The solution? Small males hang out near a dominant male and don't croak at all. When a female swims over to join the deep-voiced male, the sneaker male intercepts her, and they cuddle.

It's not clear whether the female thinks he was the one calling, whether it's a case of forced copulation, or something else, but one thing is certain: it works. Sneaker males fertilize eggs, and although it's not as effective as being a dominant male in terms of number of eggs fertilized in a season, it's a work-around that allows small frogs to pass on their DNA.

Another sneaker male strategy has been described for a frog living high in the mountains of Spain.[18] There, a sneaker male will find a cluster of eggs floating around in the water that have already been fertilized. With the parents long gone, the sneaker male grabs the floating mass of eggs using the same body posture that a male typically uses when grabbing a female. Then he releases his sperm onto the eggs. Paternity analyses have revealed that around one-quarter of the frog embryos in a pond had sneaker dads who used that strategy. Again, it's better to be a dominant male, but just because you can't play with the big kids doesn't mean you're out of the game.

Oh, we're not done with sneaker male frogs yet, though. There's a toad called *Rhinella* that is commonly found in little ponds next to streams that appear after a good rain in the Amazon.[19] When those ponds appear, hundreds of *Rhinella* toads will go there to mate, and thousands of eggs will be laid in the course of two or three days. Because that time span is so short, competition among males is ridiculously intense, and many of the females end up getting injured or killed in the fray. First a male will grab onto a female to start mating, but then another male will grab on and try to pull that first male off. Soon a female has a large ball of males holding her down, pressing her underwater until she drowns.

After the carnage is complete, animals go their separate ways, leaving several dead females lying in the pond among the fertilized eggs. At that point, a sneaker male will grab a dead female, cuddle her as he would if she were alive, and begin mating with her dead body. While he's committing this act of necrophilia, he squeezes the sides of her body, causing eggs to come out into the water one after another, like slimy pearls on a mucus necklace. Other males will sometimes approach and try to pull him off, so that first male might end up pushing her dead body around the pond, to fertilize all the eggs himself.

Incredibly, this works. Necrophiliac sneaker male *Rhinella* pass on their DNA. Males have been observed having sex with dead females in many other animal groups, including penguins, ducks, and lobsters, but *Rhinella* is the only animal for which necrophilia can actually lead to offspring.

To me, this story is an especially good reminder that nature isn't a model for what humans should do. Somewhere in a mountain pond in Spain, a male frog is squeezing eggs out of a female that was killed by gangs of aggressive, copulating males. Next time someone uses nature as a justification for any decision they've made, like what a woman's role should be in the home, perhaps, just ask them how *Rhinella* fits into their nature-inspired world.

In those instances where sneaker males mate with living females, the sex has to happen relatively quickly, or else the dominant male will have time to come over and beat the sneaker up. Marine iguanas, found only on the Galápagos Islands, are another species in which sneaker males operate, and they've found a handy way to speed things up.[20]

Female marine iguanas will mate only over a few short weeks each year, and the fierce competition among males has large males defending large territories then, for access to females. You might not know that a male marine iguana needs about three minutes of sex to achieve orgasm (though that's a great conversation starter at parties, in my experience; you're welcome). For a large male, that's no issue, but smaller, sneaker males don't have that kind of time. Anytime a sneaker male manages to have sex with a female, there's a very good chance the dominant male will come break things up before he can transfer his sperm, thus making his sneaker strategy useless.

Here's how sneaker male marine iguanas deal with that problem. When a sneaker male sees a female walking by, he gets into a sex position, with a bent body and tail, just as he would if he were having sex with her. This apparently is how iguanas masturbate, because it causes a few drops of semen to come out of his penis. That semen dries on the tip of his penis, and when he eventually does have sex, that old semen gets deposited inside the female. He'll get even more sperm inside her if he manages to stay on top of her for the full three minutes required to have an orgasm, but just in case he gets pushed away by the dominant male before that happens, some of the dried semen he produced during masturbation will be left inside her. And that's better than nothing.

Even teeny-tiny animals use sneaker male strategies. If you ever go snorkeling on a moonless night in the Caribbean, you may luck into seeing blue pulses of light shooting up through the water in complex patterns from the sea grass below. I've never seen this myself, but it's on my life list to see someday. Big-time.

Those light pulses come from special light-producing organs on shrimplike creatures called ostracods.[21] The males are tiny, less than two millimeters long, and they make those light displays to attract females. Over the course of about twelve seconds, a male flashes ten to twenty times, swimming upward through about two feet of water, leaving little blue glowing dots behind. Males make these displays in huge numbers, and I imagine they are breathtaking. Female ostracods certainly think so. A female impressed by a male will approach in darkness, then grasp the male of her choice. They have intercourse, and he places a sperm packet inside her body, before she swims off to go brood her eggs.

The problem with those displays is that they also attract predators. Lighting up and then moving in a very predictable pattern is a great way to end up in a predator's stomach, but the male ostracod really has no choice if he wants to mate.

Well, actually, there is one other option. Since a male moves in a predictable pattern, a sneaker male can position himself immediately above another male and then swim just ahead while the performer makes his display. With any luck, the sneaker male can intercept an approaching female, and she'll mate with him instead of the male who did all the work. What's different about ostracod sneaker males, compared to just about any other kind of sneaker male, is that males switch back and forth between the two mating strategies as much as they want over the course of a night. Unlike frogs and iguanas that have their roles dictated by their body sizes, ostracods seem to have a little more flexibility and to know how to use it. The ostracod embodies for me the mind-boggling wonder of the natural world. Here you have a tiny animal doing secret light-emitting dances in the night, with a mating ritual as complex as that of any bird or mammal, even though it's smaller than the head of a pin. The world is full of surprises, and the more you learn about any particular group of animals, the more surprising it becomes.

Another ocean-dwelling sneaker male that makes me smile every time I think of it is the giant Australian cuttlefish. It's an eight-legged cousin of the octopus, up to twenty inches long, weighing in excess of twenty pounds. Cuttlefish are well-known for their ability to instantly change color, which allows them to camouflage when they need to, make flashy displays to scare off predators, or communicate with other cuttlefish. And of course, if you can communicate, you can lie. Sneaker male cuttlefish are fantastic liars.

Male cuttlefish aggressively defend females, but the females aren't crazy about commitment. There are lots of males to choose from, and even though she rejects 70 percent of mating attempts, she still has sex up to seventeen times each day. About 65 percent of the time, that's with a large, dominant male. But a smaller male has three strategies for meeting females. Sometimes he'll approach a female out in the open, right in front of the dominant male (but he has to hurry); sometimes he'll approach a female under a rock, out of sight of the big male; and sometimes he'll use a third strategy—dressing in drag.[22]

You can tell a female cuttlefish from a male cuttlefish by the shape of her fourth pair of arms and by her mottled skin patterns. So "cross-dressing" sneaker males tuck their fourth arms underneath their bodies, change their body colors to look like a female, and then take on the same body posture a female takes when she's about to lay eggs. Females about to lay eggs aren't usually receptive to mating advances, so that body posture keeps most other males from trying to mate with the disguised cuttlefish.*

Pretending to be a female is a sneaker strategy that can pay off. Disguised, a sneaker male can often sidle right up alongside a true female, even while she's being guarded by a large, aggressive male. When the sneaker male makes his move toward the female, sometimes she rejects him, sometimes he gets chased off by the dominant male, but quite often, mating goes just fine.

Another male who pretends to be female is the red-cheeked salamander of the southeastern United States. When red-cheeked salamanders mate, a male and female come together and perform ritualized movements, crawling over one another and rubbing their bodies against one another in very specific ways. Being amphibians, like frogs are, male salamanders don't have penises, so during mating he glues a ball of sperm to the ground. As they move together, the female slides her body over the sperm ball and draws it up into her body.

* That said, though, the disguised male still frequently gets hit on by other males (some of whom, confusingly, are also disguised as females: it's like an elaborate Shakespearean play out there).

If a male red-cheeked salamander comes across a male and female in the middle of a dance, he will sometimes push them apart. Other times, though, he will sneak in between them and then continue the dance as though he were the female. The original male presumably doesn't realize there's been a swap, so the dance continues until he deposits his sperm packet. Then the false female turns around and bites him, causing him to run away.[23]

To me, sneaker males embody envy even better than thieving predators do. When smaller males within a species bump heads with larger males, little guys usually end up holding the short end of the stick. Whether or not he's *literally* envious is a question that might someday be answered by clever experiments, but since this happens in so many kinds of animals, we can never possibly know for all of them. Whether or not it's envy, those smaller males seem motivated to make the most of what they've got to get their genes passed into the next generation. In nature, the game's not necessarily over for a male who gets sand kicked in his face.

Now, this is exactly the kind of moment when we're tempted to look at these animals and say something about what they can teach us. We may want to point to Bill Gates and Mark Zuckerberg as nerds who came out ahead in the game of life, but that's a dangerous, slippery slope. Let's say you look at alternative mating strategies in male animals and then say something to a teenage boy like "Listen, even if you're not on the football team, you'll eventually meet a woman who loves you for who you are, who sees your sweetness, your intelligence, or your sense of humor." That's a lovely thing to say, but that's not what's happening out there. You're not asking the teenager to intercept a female on her way to a date with the football player and force her to copulate with him.

You wouldn't tell that male teenager to sneak into a couple's bed while they've having sex and take the role of the female so that the football player will ejaculate in the wrong place. You can't pretend nature's an instruction manual. It's fine to take inspiration from nature, but you still have to have common sense. Humans should abide by morals, and nature is not a place of morality. The fact that something happens in nature can never be used to justify it in human behavior, no matter how harmless it might seem to do so.

\\\\\

When my first real girlfriend and I broke up in high school, she sent me a note (this was before text messages) that said "Tell me, Dan. Is ignorance really bliss?" To this day I still have no idea what she was talking about (maybe that was her point), but for some reason, that concept has haunted me ever since. If ignorance is bliss, then *could gaining knowledge make you sadder?* If you spend your life seeking new information, is it possible that some joy might be lost?

As a parent, I'm starting to think so.

Learning about the natural world has made my life richer than it would otherwise be, but I know that learning about evolution has also peeled away some fairy tales that might have made things a little easier. I imagine that a certain bliss might have come from the belief that my great-aunt Claire could look down from heaven on the birth of Sam and smile along with me at the family resemblance. It might also have been helpful to believe in those tense moments before Sam's first breath that some benign natural force was looking after him. But the time I've spent learning about science has led me not to believe those stories.

In those instances, maybe it's possible that my science-based perspective has robbed me of a little happiness, but I decided long ago to keep my experience of the world rooted as firmly in reality as possible. That said, though, I can't help wondering what it would be like just to love Sam without that background conversation about meat robots playing in my head. It's almost enough to make me envious of those parents who don't think about evolution at all.

Almost.

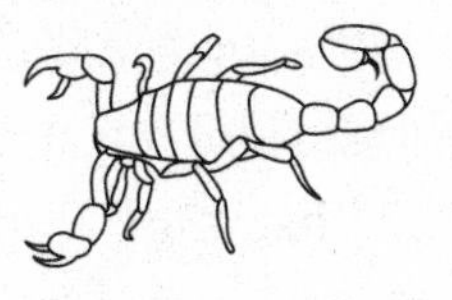

6

WRATH

Venoms and Extinctions

Nature is a violent place. Everyone knows that living things are killed by other living things all the time, and yet despite that fact, people cling to the myth that nature is peaceful. There's a make-believe world where even the predatory animals are gentle, if you just take the time to get to know them. Most of the time, believing in that fictional world is harmless, but in a worst-case scenario it can get someone killed.

The killer whale is a perfect example of a deadly animal that people underestimate all the time. Most people know what a killer whale looks like, but few have any idea what they do with their time in the wild. Instead, people watch *Free Willy*, or they go see killer whales do tricks at places like SeaWorld, and they get the overall impression that a killer whale is kind of like a family dog. In fact, at SeaWorld in Orlando, Florida, you can watch a whale named Tilikum do all kinds of tricks, just like a dog does (and it's much more impressive to see a 12,000-pound animal do what a trainer tells it to). Literally millions of people come to SeaWorld each year—many with their kids. Many of them probably believe that the whale they're looking at is Willy from the movie, or at least a friendly whale like him. But the very whale they're looking at has actually killed people. And it's happened three separate times.[1]

The first death happened in 1991, when Tilikum lived at a marine park in Victoria, Canada, with two other whales. One day, a twenty-year-old trainer accidentally fell partway into the pool. Before she could pull herself out, one of the whales snatched her by the leg and dragged her under the water. The three whales ignored the other trainers' attempts to distract them and took turns pushing the woman under as she tried to swim away, playing with her like a toy, even tearing off her clothes with their teeth, until she was dead.

Soon after that tragedy, Tilikum was moved to SeaWorld, where he still lives today. In 2010, almost nineteen years to the day after that first incident, a forty-year-old trainer was lying at the edge of the pool with her head near Tilikum's when the whale grabbed her in his mouth and pulled her under. She tried to escape but was repeatedly pulled underwater, then allowed to swim up to the surface. Then the whale pushed her around in the pool with his nose. Her jaw was broken, her spinal cord was severed, and she drowned.

Presumably those two trainers, having spent considerable time with whales, knew just how dangerous they were. The third death, though, wasn't of a trainer. It involved a member of the public, someone who may well have been living in the fictional world where killer whales are gentle giants. In the summer of 1999, midway between the two trainer incidents, a twenty-nine-year-old man visiting SeaWorld watched Tilikum perform and then hid somewhere on the SeaWorld property until the staff had closed up shop and left for the day. The next morning his naked body was found underwater, draped across Tilikum's back. His bathing suit lay at the bottom of the pool. There were no video cameras or witnesses, so no one really knows what happened to the man, but the clues hint at a gruesome death.

The first clue was that the man's clothes (other than his bathing suit) were found in a neat pile by Tilikum's pool, suggesting that he had planned on swimming with the whale. Second, when he was found, the man's body was covered in cuts and bruises, he had bite marks on his face, and his scrotum had been torn open. Those wounds show that Tilikum had played with him underwater, ripping off his swimsuit, the way he had ripped the clothes off that first trainer eight years earlier. Another clue: the man had especially deep bite wounds in one of his legs. Perhaps he had dangled a foot into the water and been pulled in by the whale, or maybe he jumped in and then tried to get out, only to be pulled back underwater. We'll never know. What is clear, though, is that Tilikum is not a gentle animal.

There's no way to know what that man was thinking when he hatched his plan. It's been suggested that perhaps he was knowingly committing suicide, but I think the much more likely explanation is that he didn't see it coming. My guess is that he believed in that imaginary world of friendly whales that theme parks perpetuate to bring in visitors. I've even played the scene out in my head a few times. It's night, and he approaches that pool, believing he's about to experience something magical. The moment of that first snatch of his leg would have been a disorienting surprise, and then the next few minutes would have been pure hell. That said, all I can really do is speculate.

In the wild, killer whales have been documented feeding on more than 140 different kinds of animals, from salmon and sharks to seagulls and sea lions.[2] Interestingly, each pod of whales will pick one kind of food and stick with it. Some eat only fish. Others, living in the same waters, will eat nothing but mammals, such as porpoises and seals. The fish eaters and the mammal eaters don't mate with one another, and they don't seem to interact at all. Perhaps in a few million years they'll be separate species, but for now they're like different societies within the whale population.* There's a lot we don't know about killer whales, but what we do know about those mammal-eating killer whales offers insights into why Tilikum may have done what he did.

When a killer whale catches a seal or dolphin, it needs to immobilize it. A whale doesn't want its meal to swim away after the first bite. Also, having an animal struggle while you try to gulp it down might cause you injuries like tooth breakage. Worse, the defensive bite of an eight-thousand-pound elephant seal can do quite a bit of damage, even to a killer whale weighing ten to twenty thousand pounds. So it's best for the predatory whale to stop its prey from moving, and one way to do that is to smash it to smithereens.

For example, killer whales often toss seals or dolphins several

* We've known since the 1970s that killer whales fit into two groups: "resident" whales, which feed on fish, and "transient" whales, which eat marine mammals (Baird and Dill 1996). Recently, though, killer whales that feed on sharks out in the open ocean have been documented. It's not yet known whether those whales form a third, separate group or whether they're just a subset of one of the other two (Ford et al. 2011).

feet out of the water with their mouths, then catch them again. Sometimes a killer whale will bat a baby seal with its powerful tail, launching the animal fifty feet or more out of the water. These kinds of games can go on for hours—far longer than is probably necessary. The mammals eaten by killer whales don't always get a quick death. By the time they're eaten, they may have open wounds, broken bones, and ruptured organs. It must be a terrible way to die.

Another killer whale strategy for immobilizing mammals is to drown them. Whales can hold an animal in their teeth underwater, or they can leap out of the water and land on top of the swimming mammal, to prevent it from breathing at the surface. That works on the really big whales—like humpbacks, gray whales, and even blue whales.[3] When a pod of killer whales hunts one of those giants, the whales first surround the bigger whale; then the killer whales take turns jumping on its back, over and over, until it becomes exhausted and can be held underwater. Next, the killer whales grab their prey by its flippers and snout and drag it underwater until it drowns. From time to time, killer whales will do this to a full-grown adult, but they usually pick on baby whales separated from their mothers.*

There's a striking similarity between the kinds of immobilizing predatory behaviors seen in wild killer whales and the behaviors Tilikum performed when those people got into the water with him. He bumped and poked them until they had broken

* Once a pod of killer whales has killed a great big whale, they have way more food than they can eat right away. In Alaska, killer whales will sometimes store the giant's carcass near a beach. The cold water acts like a refrigerator, and the killer whales can come back for leftovers repeatedly over the course of several days (Barrett-Lennard et al. 2011).

bodies and drowned. He may have been acting on instincts, or he may have remembered how to hunt from those first two or three years of his life before he was captured near Iceland and put into a pool. He didn't eat any of those humans, but it's not hard to understand why he brutalized them. He's a killer whale. That's why they're *called* killer whales.*

Killer whales, and other predators like them, hurt their prey because there's a benefit to immobilizing the prey animal, whereas there's no benefit to minimizing the prey's pain or suffering. Because evolution favors the selfish, it has endowed whales, cats, dogs, and many other intelligent predators with the instinctive desire to "play" with their prey while they kill. Because of predators like the killer whale, nature has become a place where animals wreak unimaginable pain and suffering on one another.

The tragedies surrounding Tilikum are a sobering reminder that wild animals don't treat death and torture with the disdain that humans do. The simple fact is that animals just don't hesitate to use deadly force when it benefits them, and they don't have any reason to treat the animals they kill with respect or dignity.

That kind of carnage isn't just restricted to whales. It's all over the animal world. For example, there's a North American songbird called a loggerhead shrike that impales animals alive on barbed-wire fences like trophies. To look at the bird, you'd never expect such brutal behavior: it's just a little black-and-white thing, a bit smaller than an American robin, with a tiny, barely perceptible hook at the tip of its beak. But that small size is precisely the

* I was once told on a whale-watching tour that "killer whale" is no longer the accepted name for these animals, and that the term "orca" is preferred, to reflect the fact that wild killer whales/orcas don't hunt humans. I think the name suits them beautifully. (Not everything has to be about humans.)

reason the shrike has to do such morbid things to the animals it eats.

For their size, shrikes eat large animals—large insects, lizards and snakes, even birds and mice, some of which weigh half as much as the shrike itself. A prey animal that size is going to put up a fight if you try to eat it, and a shrike doesn't have powerful raptorlike talons to subdue those animals while they rip them apart. That's where the barbed-wire fence comes in. By impaling its prey alive on a meat hook, the shrike can leisurely tear into its body with that hooked beak without having to hold the animal in place. Sometimes shrikes use the thorns of plants as hooks, but when barbed-wire fences are available, that seems to be what the shrikes prefer.

Shrikes of both sexes use impaling behavior to help them feed, but males also use impaled prey to mark their territories and to advertise to females how good they are at hunting.[4] I don't know what's creepier—that males perform that behavior or that females are turned on by it. Either way, a barbed-wire fence with dead animals spaced out at even intervals on the hooks is reminiscent of severed human heads lining the fence of a medieval fort. If the shrike version weren't so small, it would be terrifying.

Killer whales and shrikes couldn't be less similar in terms of body size, where they live, and what they eat, but as predators they're united by the need to immobilize their prey, and both accomplish that task by overpowering them. That's also true of many other kinds of predators, from lions to crocodiles, from eagles to great white sharks, and from mongooses to wolves. Those predators have to be stronger, have better acceleration, or have better stamina than the animals they eat. They feed because they can beat their prey in contests of brute-force physics.

But not all predators can win those physical contests, so a discussion of wrath in the natural world would be incomplete if I stopped there. Even if a predator isn't strong enough to overpower its prey with physics, it might still have a chance, using chemistry.

Venom changes everything.

Some people use the terms *venom* and *poison* interchangeably, but there's an important difference between them. Poisons are chemicals that animals use to prevent other creatures from eating them. Think of the poisonous skin secretions of a poison dart frog, for example. Venom, on the other hand, is a cocktail of chemicals that is injected into a victim with the specific purpose of harming it. Venom can be used for offense or defense; sometimes it's used to subdue prey, and other times it's used for protection from predators. In either case, venom is a chemical weapon. (Next time someone tells you about a poisonous snake, feel free to correct them. Snakes are sometimes venomous but never poisonous.)

Using chemistry instead of physics to hunt prey means you can eat things that are much stronger than you are, and I can think of no better example than that of a jellyfish killing and eating a fish. It's only thanks to venom that a boneless blob is able to take down a muscular animal that normally swims circles around it.

Jellyfish deliver their venom through microscopic harpoons. Hundreds of thousands of them line each tentacle, and when a fish swims into a tentacle, the harpoons fire into its flesh, squirt venom, and then stick into the fish's skin with back-pointing spikes, just like real harpoons do. This happens unbelievably quickly—the harpoon tips accelerate with a g-force 40,000 times greater than gravity, and once you touch a jellyfish stinger, the whole thing is over in less than three milliseconds—a fortieth of the time it takes you to blink.[5] Jellyfish venom quickly immobilizes the small animals that jellyfish eat, but it also works as a defense mechanism, preventing some animals from eating the jellyfish and causing other swimming animals, like us, to simply give the jellyfish all the space it wants.

For humans, the number one jellyfish to watch out for is the Australian box jellyfish. It weighs more than two pounds and has up to sixty ribbonlike, six-foot-long tentacles. If you swim into those tentacles and enough of those microscopic harpoons dump their venom into you, you'll get excruciatingly painful lesions all over your skin. But that's just what happens on the *outside* of your body. On the *inside*, things can get much, much worse.

The Australian box jellyfish's venom makes potassium leak out of your red blood cells. As the potassium leaves, so do water and other molecules, like hemoglobin, causing the cells to become deflated and limp. Eventually, the red blood cells break apart and dissolve into fragments. That's bad because you need red blood cells to carry oxygen from your lungs to the tissues all over your body. If the red blood cells are all destroyed, tissues all over your body will start to suffocate and die. That's not what does you in, though. You see, all that leaked potassium, now floating around in the blood outside the red blood cells, starts to interfere with the chemical mechanisms that make your heart squeeze properly during a heartbeat. That's how the sting of an Australian box jellyfish can cause your heart to stop beating, and that is what kills you.[6]

The Australian box jellyfish is easily the most deadly jellyfish to humans, but the vast majority of jellyfish species aren't dangerous to humans at all. Sometimes a jelly's harpoons are simply too small to make it through our skin. Other times, the chemical structure of the venom just doesn't have a lethal effect on the human body. For instance, there's a huge jelly you might have heard of called a Portuguese man-of-war. This jelly can have up to thirty harpoon-loaded tentacles, each about a foot long, that can quickly kill a fish. But despite urban myths to the contrary, the Portuguese man-of-war is extremely unlikely to kill a human (even though humans will feel a world of pain if they are stung).*

There's another animal out there that is even less susceptible to the stings of a Portuguese man-of-war than humans are, even though it's just a tiny thing, only about an inch long. In fact, what it can do with the jelly's stingers might be one of the most unbelievable tricks in all of biology. Back in the chapter on gluttony we talked about the emerald sea slug, which gets its energy from the sun by eating algae and moving the solar-powered molecular machinery to the surface of its own skin. Taking a page from the same book, there's another kind of sea slug, called the blue dragon, that eats the stinging tentacles of a Portuguese man-of-war without setting off the harpoons. Then, like its solar-powered cousin, the blue dragon moves those harpoons to its own skin for protection.[7]

The Latin name for this stinging sea slug is *Glaucus,* after the Greek sea god who was born a mortal but became immortal when he ate a magical herb. What a perfect name for such an unusual creature. It avoids getting stung itself by covering its skin with

* The Portuguese man-of-war gets its name from the boatlike air-filled float that sits on the surface of the water, above the dangling tentacles (a man-of-war was a type of warship used in the sixteenth to nineteenth centuries). That float allows the jelly to be moved around by wind, in addition to ocean currents (Šuput 2009).

a mucus that chemically inhibits the man-of-war harpoons from firing. The blue dragon also has specialized skin cells that can absorb any harpoons that do fire, almost like sandbags, preventing the venom from leaking into its body. Since the blue dragon eats the harpoons, it has similar linings in its mouth and all along its digestive tract as well. Once the harpoons are ingested, they're taken up by specialized mover cells and then placed out at the skin of the animal. There they are kept alive by secretions from the blue dragon's body, ready to fire if the blue dragon needs protection.[8]

While that particular sea slug has to steal venom from jellies, it has some close relatives called cone snails that can make venom on their own. You might think of snails as slow, harmless creatures, but those are not words you can use to describe cone snails. Cone snails live in tropical oceans, where they feed on fish, crustaceans, and other prey you would think are too fast for a snail to catch. But they absolutely can catch those prey, and they do so by firing a high-speed venom-tipped harpoon into the victim as it passes by. Cone snail venom is *extremely* deadly to humans, possibly the deadliest venom out there. In fact, a cone snail can kill a person in less than five minutes. Some people call them cigarette snails, because once you've been stung, you only have enough time for a last cigarette.*

Cone snail shells are, well, cone-shaped (hence the name), with a long slit along the side, out of which the animal emerges. They tend to be brightly colored with intricate patterns, making them absolutely beautiful. Once you know what they look like,

* Obviously I don't condone smoking, but if you've just been stung by a cone snail, an increased likelihood of lung cancer is the least of your problems.

they're unmistakable, so I can only assume that most of the people who have been killed by cone snails picked them up for their beauty, with no idea what they were. Or perhaps some of those victims did know the dangers of cone snails but assumed that their wetsuit gloves would protect them from the harpoons. But unlike the millions of microscopic harpoons that jellyfish have, the much larger, single harpoon of the cone snail is a modified tooth and easily large enough to penetrate a wetsuit.

Interestingly, the venom of a cone snail is almost painless because it contains powerful proteins that block pain.[9] That makes them attractive animals of study for pharmaceutical companies: perhaps cone snails will someday reveal nonaddictive alternatives to drugs like morphine. That said, finding the pain-blocking proteins in cone snail venom is a bit of a needle-in-a-haystack problem. Each species of cone snail produces more than a thousand different proteins as part of its venom, and those proteins vary among the five hundred or so species of cone snail. In fact, the proteins can even change over the lifetime of an individual snail.*

Cone snails rarely kill humans because most humans don't spend very much time underwater in the tropics at night, and those who do usually know enough about cone snails to stay away from them. That's why, despite their extraordinarily deadly venom, cone snails have only been blamed for the deaths of around thirty people.[10] Instead, people are much more likely to be envenomed by creatures that live where humans do, back up on land—things like spiders.

There are more than 44,000 species of spiders in the world

* Only about 0.1 percent of those proteins have been analyzed well enough that we have any clue what they do (Lewis et al. 2012).

and almost every one of them is venomous.* Most don't hurt humans though, either because they don't bite in defense, because their fangs can't pierce human skin, because they produce such a small amount of venom, or because their venom just doesn't happen to react chemically with the human body. Of those spiders that do cause problems for people, most only cause short-term pain and swelling right at the site of the bite. For the most part, despite their abysmal popularity ratings, spiders aren't really anything to worry about at all. That said, there are a few out there that can do considerable damage, and spiders from a few select groups have actually killed people. Overall, though, deaths from spider bites are vanishingly rare. Even the bites of spider species that *can* kill humans are almost always nonlethal.†

Spiders deliver venom through stingers on either side of the mouth, and most spider venoms are neurotoxins. In other words, they're chemicals that affect neurons. Neurons are the cells in your brain that enable you to think, and they also run down your spinal cord and out into nerves branching throughout your body. Insects have neurons too, and it's usually those insect neurons that spider venoms target, since most spiders eat insects. Whether or not those venoms, intended for an insect's neurons, will have any effect on a human is mostly a matter of luck. Some spider toxins have no effect at all on humans. But others work *very* well.

---

* An arachnologist (spider expert) named Norman Platnick at the American Museum of Natural History, in New York, keeps an online database of all the spiders that are known to science. Last time I checked, he had 44,032 species listed (Platnick 2013), but the list of spider species that haven't been described yet may be four times that long (Vetter and Isbister 2008).

† Specifically, the deadly spiders are the Australian funnel-web spiders (*Atrax* and *Hadronyche*), the South American armed spiders (also called banana spiders, *Phoneutria*), and the recluse spiders (*Loxosceles*) and widow spiders (*Latrodectus*) found across several continents (Isbister and Fan, 2011).

A neuron is a long cell with branches at either end that interact with the branches of other neurons. A message travels along the length of a neuron until it reaches the end. To continue, the message must be passed on to another neuron. The first neuron passes on its message by releasing chemicals called neurotransmitters from the tips of its branches. Those neurotransmitters float across the microscopic space separating the branches of the first neuron from the branches of the next, and once they've made it across (a process that takes only milliseconds because the distance is so short), the receiving neuron fires, sending that electrical wave down the length of its own body, on toward the next neuron in line. This whole process is unimaginably fast and can obviously only happen if the neurons are working properly.

Many spider venoms affect the neurons of their prey by throwing off the ability of neurons to control their electrical charges. That means that when the neuron receives a message to fire from the neuron before it, it short-circuits. Different venoms accomplish this in different ways. Some cause the neurons to leak potassium, others sodium, and still others calcium. In their own way, each of these can make a cell go haywire. What's worse, the venom of any particular spider typically contains many different neurotoxins, each of which messes up neurons in slightly different ways.

In addition, spider venoms often also contain other chemicals that cause swelling and pain.[11] Unlike the neurotoxins that are intended for insects, those other chemicals make the venom especially effective for protection against larger animals that could hurt the spider—animals like us.

The most deadly spider in the world to humans is a half-inch to two-inch black beast with massive fangs called the Sydney funnel-web spider. There are other funnel-web spiders in other parts of the world, but none of those are very dangerous at all. Sydney funnel-web spiders, though, are very different. They're aggressive, they live right around a city with a population of 4.5 million people (Sydney, Australia, in case you hadn't guessed), and their venom contains a particularly nasty neurotoxin that messes with the neuron's ability to control sodium movement in and out of the cell. For whatever reason, the venom of that particular spider is especially hard on humans. In fact, the venom of the Sydney funnel-web spider doesn't really do anything to cats, dogs, rats, or other nonprimate mammals. It's just really effective on primates, even though primates aren't native to Australia.[12]

An antivenin now exists for the venom of the Sydney funnel-web spider.* It's created by taking the venom out of several Sydney funnel-web spiders, then injecting them into a rabbit, at first in small doses, and then in larger and larger ones. Over time, the rabbit builds antibodies to the venom, until eventually its blood is full of antibodies specifically made to attack the neurotoxins in that spider's venom. Then the rabbit's blood serum (blood without the red blood cells, basically) is drawn and stored in vials. Later, when someone gets bitten by a Sydney funnel-web spider, the person is injected with antibody-laden rabbit serum. That gives the human's immune system some help breaking down the neurotoxins, rendering the venom of the spider nonfatal. It's a great trick, and since the antivenin was created, in 1980, there hasn't been a single fatality from the Sydney funnel-web spider (knock on wood).[13]

* Apparently, *antivenin* is the correct term for what should really be called an *antivenom*. I know. It drives me nuts too, but if you want to sound like you know what you're talking about, you have to say "antivenin."

A lot of people find spiders creepy, but if there's a creepy crawly out there that makes people even more nervous, it has to be the scorpion. Scorpions are eight-legged arachnids, like spiders (as opposed to six-legged insects). Like spiders, scorpions also have venom, but instead of delivering it through the mouthparts, scorpions, as everyone knows, sting you with their fabulous tails.*

Scorpions have existed, mostly unchanged, for more than 400 million years, since well before the dinosaurs and before other animals had yet come on land, including insects and our own ancestors for that matter. The biggest scorpions alive today are just over eight inches long, but millions of years ago there were three-foot-long scorpions walking around, which have since been given outstanding names, like *Gigantoscorpio* and *Brontoscorpio*. (With names like those, it's hard to believe they haven't yet been featured in some dreadful made-for-TV horror movie.) In all likelihood those animals lived in the water, where buoyancy would have helped them support their enormous bodies. But still, it's sobering to imagine coming across one.[14]

There are more than 1,500 scorpion species out there, and of those, 25 or so have a sting that can kill a human. Despite that small number, roughly five thousand people around the world die each year due to scorpion stings, making bees and snakes the only venomous animals that kill more humans.[15]

The first time I ever came across a scorpion myself was in

* The class Arachnida includes spiders (order Araneae), scorpions (order Scorpiones), mites and ticks (order Acari), and a bunch of other, less well-known eight-legged creatures. Personally, I'm very creeped out by tailless whip scorpions (order Amblypygi) and the so-called camel spiders (order Solifugae). I think it's probably because the first time I ever encountered those animals I didn't even know they existed, and because both of them have mouthparts that look like they were thought up by David Cronenberg.

Belize, on one of those field excursions for my master's degree. I was lying in a hammock under a thatched roof when someone pointed out a scorpion crawling on the wall above my head. I leapt up, grabbed a pair of long sticks, and then used them like chopsticks to move the scorpion delicately out into the sunshine so I could take photos of it. As I shot the pictures, it first went into a great defensive pose, holding its tail and claws up perfectly for the photos, but then it suddenly went limp and died. I had no idea what killed the scorpion, but I carried it over (with the sticks—I was still too scared to touch it) to show it to the other biologists there. One of them picked it up to show me some of the anatomy on its belly, and then as we ate lunch the scorpion got passed around the table for everyone to see. Later, I was carrying the scorpion in my hand back to my cabin, trying to decide whether I should attempt to bring this thing home through customs, when it suddenly leapt back to life and ran up my arm. I screamed—*screamed*—and swatted the scorpion into the air. It landed on the ground, then walked nonchalantly away like nothing had happened. I wasn't stung, but now that I know a scorpion can play dead for three hours at a time, I know not to trust them ever again.

Like spider venoms, many scorpion venoms are neurotoxins. So even if you don't die, a sting from the wrong scorpion can cause paralysis, convulsions, and heart palpitations. Like most venomous creatures, scorpions use their venom primarily to subdue prey while hunting but also make use of it to defend themselves against predators. Most of the time that works really well, but unfortunately for the scorpions, venom doesn't work against all the predators all the time. For proof of that, look no further than the pallid bat, which specializes in eating scorpions.

The pallid bat is a gorgeous yellowish-white bat from the deserts of western North America. It weighs about as much as a sparrow, though it seems bigger when you see it in flight. The first things you notice about a pallid bat are its big, beautiful ears. When you look more closely, though, you'll see that it also has intricate, scrolling nostrils that give it unique appearance, even among bats. When it hunts, the pallid bat either sits on a perch or flies silently, listening carefully for the footsteps of large insects or arachnids. When it hears those sounds, it hovers over the spot to get a closer "look" with its echolocation, then flies down to kill its prey.[16] If it's a scorpion, the first thing the bat does is rip off the stinger. Sometimes, though, the bat's just not fast enough and gets stung anyway. There's good evidence that scorpion stings do hurt the bats: pallid bats fed nothing but scorpions for several days will go after them with gusto on day one but seem to lose their enthusiasm as the days wear on. Some researchers have even caught pallid bats that were missing an eye, presumably from a well-placed scorpion sting. Scorpions are a tough food to eat, but pallid bats are hard-core.[17]

I've seen pallid bats on several occasions, mostly in the cracks under bridges in Texas and California, but my most wonderful memories of pallid bats come from a night with Shelby at a campground in Texas.

A few years after I started dating Shelby, we took a road trip around Texas with my mom, to see some bats and caves. My mom's always been supportive of my weird fascination with bats, and one year she told me she wanted to see some bats firsthand. Texas was an obvious destination, since it has those spectacular bat caves we discussed in the chapter on gluttony, filled with so many Mexican free-tailed bats that it takes up to four hours for all of them to come out each night. My favorite of them is the Eckert James River Bat Cave Preserve near Mason, so my mom and Shelby and I started there.

Once we'd seen that bat cave, I told them about a bridge four hours away, under which pallid bats had been seen about ten years earlier. On the off chance they might still be there, we decided to make the drive.

Unfortunately, the bats weren't under the bridge (although we did find a complete cougar skeleton lying on its side there, which was exciting). That afternoon, we drove to the nearby campground at Balmorhea State Park, set up our tents, and had dinner. As the sun set, my mom went to sleep in her tent, and Shelby and I went out for a late-evening walk to look for wildlife. I love southern Texas because you can see all kinds of things there, from tarantulas to giant walking sticks, that are absolutely nothing like the animals I grew up with in Edmonton. That night we even lucked into seeing a scorpion. What with the cougar skeleton, the scorpion, and the general beauty of the area, we'd had about as good a day as a couple of biologists could possibly have. It was about to get way, way better, though.

As it got darker, Shelby and I kept walking until it got dark, eventually arriving at a playground on the edge of the campground, where we stopped to sit down in the sand. Then, using a trick my PhD advisor had once taught me (he had done part of his own PhD on pallid bats), I scratched the sand with my finger. Apparently, that makes a sound very similar to the sound a scorpion makes when it walks in the sand, and to a pallid bat, that's practically a dinner bell.

Almost instantly, there was a bat hovering right in front of my face.

I couldn't believe it.

I couldn't turn on my headlamp to look at it, even though it was almost pitch-black out, because I knew that would probably scare it off, so I let my eyes adjust to the darkness and tried to make out its shape. I could clearly hear its flapping wings but could barely see it. It hovered for several seconds, but then, as suddenly as it had appeared, it flew off.

Immediately, I tried again, and sure enough, another bat showed up. At first, Shelby and I whispered back and forth with the requisite "Oh my *God*!" and "Can you believe this is really happening?" but before long we just went silent to enjoy the experience together. We sat together at first, but then lay down in the sand, holding hands, looking up at the stars, and calling pallid bats in with our fingertips. I don't know whether just one bat found us interesting or if we fooled an entire colony, one after another, into believing there were scorpions in the sand. But whatever the number of bats we saw, it was a perfect night.

In fact, that dark night epitomizes what I love most about nature. There I was, outside, with the woman I love, watching silhouettes of my favorite creatures against a backdrop of stars. On the one hand it was totally peaceful—the silence, the cooling wind, the fresh air—but on the other hand I felt defenseless and vulnerable, lying on my back in darkness surrounded by insects, scorpions, and, for all I knew, more cougars. But the way those elements took me out of my comfort zone is a big part of what made that night so special. To me those scary animals are what bring nature to life. Knowing I'm surrounded by animals that could hurt me reminds me that nature's not weak or passive. It reminds me to respect the natural world.

Unlike many experiences in nature I've had alone, that night was also a special one for Shelby and me as a couple. It was wonderful for both of us to be able to share an experience like that, with the hostility of the creatures in that environment juxtaposed with the kind of relationship the two of us had built. Against a backdrop of wrath, the caring, loving, and patient affection Shelby and I share was even more apparent.

From an aesthetic perspective, nature's wrath is a big part of what I think makes it so beautiful, but my poetic thoughts don't mean crap to any of the animals out there. On a day-to-day basis, that wrath exists because it can spell success for the animals that wield it. One Texas resident, the venomous fire ant, is a perfect example.* I weigh around 50 million times what a two-milligram fire ant does, but a fire ant can hold its own against me just fine. The bite of a single individual can leave a welt on my skin that lasts for days, but that's not what scares me about them. It's the fact that a person never gets *one* fire ant bite. Fire ants, like most ants, are successful because they deliver their wrath en masse.

When you accidentally stand on top of a fire ant colony in your sandals, they come at you right away, but they don't bite you in those first moments. Instead, they invade enemy territory—crawling up your legs by the hundreds. Because of their small size, they're barely noticeable. But you notice them soon enough, because eventually one of those ants does bite you, and in so doing, releases a smelly chemical into the air that gives everyone else the signal to bite too. As nearby ants bite, they release more of that

* Fire ants aren't actually native to the southern USA. They're from South America but got introduced via cargo ship to Mobile, Alabama, in the 1930s. Today, fire ants span the southeastern USA and Puerto Rico, and there's a very real possibility that their range will continue to spread until it covers the entire southern USA, right across from Florida to California.

same stinky chemical, causing others to bite, and so on. Not only that, each ant can bite multiple times. As a result, one minute you feel nothing, and then suddenly every square inch of your legs burns at once. If they've gotten far enough up your clothes, you'll need to drop your pants immediately to get the ants out of there. Despite how socially awkward that might be, you really have no choice.*

Once the ants are gone, the bites are just a little itchy and annoying, but over the next few days they grow, until a day or two later the pus-filled welts on your skin are even bigger than the ants that left them.

Fire ants suck.

Ant bites are painful, partly because of an acid that ants have in their venom. It's called formic acid, after the Latin word for ant, *formica*. Working as teams, injecting their enemies with formic acid, ants have flourished around the world. There are more than 12,000 ant species out there,† and the number of individual ants walking around on Earth is almost unfathomable. Biologists have famously estimated that if you took every single land animal out of the Amazon rainforest and put them all on a scale at the same time, about one-third of that weight would be made up of ants and termites.‡

One of the Amazonian ant species that would be on that scale

* As a German colleague of mine once said outside a bat cave in Texas, right before she pulled her pants off in front of our whole lab group, "It is like hell!"

† Last time I checked, there were 12,763 ant species known, but it's likely at least that many are still out there waiting to be discovered. A running tally is kept by scientists at Ohio State University: http://osuc.biosci.ohio-state.edu/hymenoptera/tsa.sppcount?the_taxon=Formicidae.

‡ I know that seems like a bit of a weird thing to think about, but it really does drive home just how many ants are out there (Fittkau and Klinge 1973).

is called the bullet ant. I don't know if it got that name because it's an inch long, roughly the size of a bullet, or because when it stings you it feels like being shot. Either way, when Shelby was doing fieldwork for her PhD in the Brazilian Amazon, the locals often warned her about bullet ants. They had a different name for them, though: *vinte e quatro*, or "twenty-four"—apparently the number of hours for which you can expect to be in searing pain after being stung by one.

There's an indigenous tribe in the Amazon, called the Sateré-Mawé, that uses bullet ants in a rite of passage for its men. During a ceremony called the *tucandeira* ant ritual, the young man puts his hands into a pair of ceremonial gloves that look like giant oven mitts woven out of leaves. Each glove is filled with up to three hundred bullet ants, and the man must leave his hands in there for several minutes while he does a ritual dance. The thing with bullet ants is that the pain gets steadily worse after the stinging and climaxes several hours later, so the ritual goes on for hours after the men take their hands back out of the gloves. This ritual starts for men around the age of twelve and may be repeated twenty-five times or more in a man's life.*

The sting of the bullet ant has a pretty solid reputation as the most painful of any insect, but of course the only way you could really know that would be to go out and get stung by as many different kinds of insects as possible. And if you think no one would ever do something like that, you clearly haven't spent enough time with entomologists.

Justin Schmidt studies ants and their close relatives, the

* Indigenous tribes on the outskirts of Manaus, in the heart of the Amazon, make money by performing this ritual in front of tourists. Personally, I think tourists should have to perform the ritual instead (Botelho and Weigel 2011).

wasps and bees. Collectively, these three groups of insects are called hymenopterans, and if there's one thing ants, wasps, and bees are famous for as a group, it's that they sting. In the course of studying these insects, Dr. Schmidt has been stung many times, but instead of just swearing loudly about it, he's recorded what each of those stings felt like. The result of his work is the Schmidt Pain Index.

The Schmidt Pain Index is a list of scores for different hymenopteran stings, describing how painful they are. It goes from one to four, with four being the most painful. The best part about the Schmidt Pain Index is that some of the stings are described almost the way one might describe a wine. For example, the sweat bee, with a score of one, is "light, ephemeral, almost fruity. A tiny spark has singed a single hair on your arm." The bull's horn acacia ant, which you may recall from the chapter on gluttony, scores a two and gives "a rare, piercing, elevated sort of pain. Someone has fired a staple into your cheek." There are a few hymenopterans with a score of four, and the bullet ant is one of them, described as "Pure, intense, brilliant pain. Like walking over flaming charcoal with a 3-inch nail in your heel."*

In interviews about his list, Schmidt has said that the pain of a bullet ant is the worst of them all, but although the Schmidt Pain Index contains an impressive 78 species, it falls slightly short of covering all 117,000 species of hymenopterans out there.[18] So while the bullet ant may well have the most painful bite of any insect, there's still clearly a world of pain available to any other entomologists looking to have an index named after them. A good

* Dr. Schmidt isn't some nut. He is a well-respected scientist, and I would argue that his willingness to experience those stings firsthand shows just how passionate he is about his subject of study. It's a good thing he doesn't study cone snails, though.

starting point might be the enormous, one-and-a-half-inch-long ant *Dinoponera gigantea* of Brazil, which is not part of the Schmidt list, and whose sting might be even more painful than that of the bullet ant. A medical report from 2005 of a sixty-four-year-old man stung by a *Dinoponera* ant describes incredible pain (more intense than passing a kidney stone, he said), cold sweating and nausea, a vomiting episode, and an irregular heartbeat. Three hours later, the man passed stool with a large amount of blood in it, having not had any such problems prior to his sting. His intense pain only started to decrease eight hours after he was stung.[19] I'm thinking this ant might give a bullet ant a run for its money (but I'm not going to be the person who signs up to compare them).

As impressive as it is that an ant can make you poop blood, there aren't any hymenopterans with a venom so powerful that a single sting can be lethal. Some people have allergies to fire ants or honeybees, so complications from those stings can be deadly, but if you don't count allergic reactions, or swarms of hymenopterans attacking together, the stings of bees, ants, and wasps are mostly just painful. In fact, adding up the jellies, the cone snails, the spiders, the scorpions, and the hymenopterans, you don't get anywhere close to the number of human deaths that you get from the top killers. None of the venomous creatures I've listed in this chapter so far has even made a slight dent in human populations. That changes, though, when you consider the deadliest of all the venomous animals out there: the snakes.

Of the 3,400 or so snake species in the world, only a few hundred are venomous, but a large number of those venomous snakes eat mammals.[20] Even though we're not the mammals that those venomous snakes feed upon, we're similar enough to the mammals they do eat that snake venoms are often deadly to us too. Combine that with the fact that snakes hide well and strike quickly, and snakes start to look like the poster creatures for this book.

It's kind of shocking how many people get killed by venomous snakes. Most deaths happen in poor countries, where people work in agricultural fields without proper footwear, far from sufficient medical infrastructure. Exact numbers are hard to come by because records in those regions are not as well kept as in other parts of the world. To give you some idea of numbers, though, the country of Bangladesh alone suffers roughly 6,000 deaths per year because of snake bites. Worldwide, the estimate ranges anywhere from 20,000 to 125,000 deaths per year.[21]

As with other venomous animals, different kinds of snake venoms work in different ways, and any given snake's venom is a cocktail of different kinds of molecules.

Some snake venoms simply kill the cells with which they come in contact, causing swelling and blister formation and ultimately causing the flesh to turn black and die. Imagine one whole side of your leg, from your ankle to your hip, with the skin and underlying flesh eaten away chemically. That's what a bite from a spitting cobra can do. And just in case that doesn't hurt enough, the bite also contains chemicals that make your body feel intense pain at the site of the injury.[22]

Other snake venoms affect the cardiovascular system. Some cause a sudden drop in blood pressure, while others do exactly the reverse, squeezing the arteries around the heart until it can't beat properly for lack of oxygen. Others affect the blood itself. They might prevent blood from clotting, as is the case with the venom of the copperhead, or they might make the blood inside you clot up into blood Jell-O, as does the venom of the fer-de-lance.[23]

Then there are the snake venoms that are neurotoxins. A few pages ago, when we were talking about spider venoms, I mentioned chemicals in your body called neurotransmitters, which your neurons use to send information to other neurons. There are two different ways a snake neurotoxin can screw that process up, but some of them are permanent and others are temporary. This difference is pretty simple: it depends on whether the neurotoxin affects the neurons sending the neurotransmitters or the neurons receiving them.[24]

Some snake neurotoxic venoms disrupt the "sending" neurons by destroying the parts of the cell that send neurotransmitter molecules. The damage caused by this kind of neurotoxin is permanent because the neurons never recover. This is what makes snakes like taipans and pit vipers so deadly. Other snake venoms interfere with the "receiving" neurons by sitting on top of their receptors, physically blocking any neurotransmitters that arrive from other neurons. This second type of neurotoxin lasts only until the venom is broken down and moved out of the way by the human's immune system. Once it's gone, the neurons can talk to one another again. That means that if a person has been bitten by the right kind of snake, a medical team that knows what to do can bring the victim back from death's door.

For example, the bite of a Malayan krait contains neurotoxins that cause paralysis of the whole body. First the person loses the ability to move voluntary muscles, but soon even the muscles around the lungs stop working. That means air is no longer coming into the lungs, so it doesn't take long for the victim to die from a lack of oxygen. However, because Malayan krait neurotoxin is of that second, reversible type, there is hope for a victim who is lucky enough to get adequate medical care. If air is pumped into that person's lungs while they're paralyzed, there's never a shortage of oxygen, and no permanent damage occurs. So long as the medical staff keeps the patient "breathing," that patient can fully recover from the total paralysis of a krait bite in just a few days.[25]

Because snake venoms work so well on humans, it's easy to forget that they really aren't there to hurt us. Snake venoms are primarily there to help snakes subdue their prey. After all, these are animals that don't have any arms and that hunt other animals that can usually move much more quickly than they can. It's pretty incredible that an animal like a rattlesnake can catch something as fast as a chipmunk at all. Fortunately for the rattlesnake, though, it only has to hunt successfully a few times each year.*

When hunting, a rattlesnake relies on its excellent sense of smell to find a spot in the forest that small mammals, like chipmunks, have been using as a trail. It would not be very easy for you or me to smell the residue on a branch from the soles of chipmunk feet that passed by four hours ago, but that kind of thing is right up a rattlesnake's alley. In fact, that's why they have forked tongues. The tongue comes out, touches different places in the forest, picking up chemicals that might be too heavy to float freely in the air, and then comes back into the mouth and touches

* One of my closest friends during my PhD studies was a guy named Rulon who studied timber rattlesnakes in the wilds of upstate New York. When he would tell people about his research, they would often ask him "What are they good for?" So Rulon wrote a piece for *Natural History* magazine that answered that question beautifully. If you want a sense of what makes those snakes such beautiful creatures, and want to see just how passionate a researcher can be about his research animals, I highly recommend that article (Clark 2005).

sense organs on the roof of the mouth. It's kind of like when you go through security at the airport and they swab your bag with that little cloth and then put it into the chemical analyzer. Once the snake detects the chemical residues of mammal footprints, it sets up shop, coils, and waits.

The wait may last multiple nights, but snakes can be patient. They're cold-blooded, so they use energy slowly. When a chipmunk does finally pass by, the snake strikes out, stabs the small mammal with its hollow fangs, injects it with venom, and then immediately lets the animal go. After all, there's no sense holding on to a struggling chipmunk, since it might bite in self-defense. If the strike was successful, the chipmunk will be dead in just a couple of minutes anyway. As the chipmunk runs away, the snake slowly uncoils and then follows the scent of fresh footprints. Before long, the snake reaches the dead chipmunk at the end of that scent trail and sets to work eating it whole. None of that would be possible without venom. In fact, the venom even acts as a digestive juice, starting the breakdown of the chipmunk even before the snake eats it, easing the digestion that will take the snake several weeks to complete.[26]

When a rattlesnake's venom enters a chipmunk, it first goes to work on the blood vessels. Within moments, capillaries and small veins near the bite start leaking blood uncontrollably into the surrounding tissue. This results in swelling, blistering, and overall breakdown of the tissue to its chemical components. As the blood carries the venom through the bloodstream, those problems spread to other parts of the body, including the heart.

Should a human be unlucky enough to step over a log and

onto a coiled rattlesnake, the animal might strike in self-defense. The potentially good news is that roughly half of rattlesnake defensive bites on humans are "dry bites," which means that no venom is injected. Dry bites still hurt because a pair of daggers just pierced your skin, but at least you won't suffer the chemical onslaught that eats away at your muscle tissues. Rattlesnakes sometimes deliver those dry bites because they're just trying to get the human to go away. Whether or not they kill the human doesn't really matter to the snake, so with all things being equal, it's better to save venom for when it's needed.*

To me, the scariest thing about being bitten by a snake is that you can't tell at first if a snake bite was a dry one or not. The only way to find out is to wait for symptoms. If you're lucky, the bite wounds swell up and become painful like any pair of deep puncture wounds would. But if you're not so lucky, well, what happens next depends on which snake bit you, how much venom it delivered, and whether you have access to antivenin. If it is a rattlesnake, and you're unlucky enough to have received a "wet bite," you'll probably still survive, though. The rapid tissue destruction will do permanent damage to the skin and muscles around the bite site, but rattlesnake venom happens to be nonlethal to humans . . . at least most of the time.

I understand that people want to believe that Mother Nature is looking after them, but how can they reconcile that idea with

---

* My rattlesnake-expert friend Rulon once told me over beer that he thought dry bites might be the reason there are so many strange folk remedies for snake bites. After all, if only half of snake bites involve venom, then any remedy will appear to have saved you 50 percent of the time. That might explain some of the folk remedies that are out there, like sucking the bite hole, soaking the bite in kerosene, killing the snake and rubbing its body on the wound, or killing a chicken and wrapping its meat around the wound.

the existence of jellyfish, cone snails, spiders, scorpions, bees, ants, wasps, and snakes? That's not even starting on the centipedes, stingrays, scorpion fish, lizards, platypuses, shrews, beetles, and bugs that also use venom. If Mother Nature is so loving, why are there so many venomous creatures out there, and why do so many venoms contain chemicals that do nothing other than increase the pain of the victim?

Some would argue that venoms provide many useful chemicals to scientists trying to develop drugs, and that's totally true, but I think it's a stretch to pretend that's why nature put them there. The fact is, we humans have learned to take advantage of nature in order to thrive as a species. Just as we eat plants and animals that didn't volunteer for the job, we take apart the venoms that animals use against us. Through science, we can make antivenins, rendering those venoms less deadly, but even more amazingly, we can use venoms to gain access to parts of our own bodies we couldn't reach before. If you're making a drug that blocks pain, see what cone snail venoms are doing. If you're making a drug that can temporarily block nerves, why not look at snake venoms? Venoms provide keys to the locks on millions of molecular machines within our bodies. With venoms we are making drugs that help us live long and healthy lives. Nature isn't taking care of us. She's trying to kill us, and we're taking care of ourselves.

\\\\\

Animals kill other animals, and there's no question that they often inflict pain when they do it. Whether it's physical abuse or chemical abuse, animals give one another a beating out there every single day. But wrath in the natural world doesn't just happen on the scale of injured, tortured, or even killed individuals. Wrath happens on much larger scales than that. The wrath of nature is strong enough to wipe out entire species at a time, or even groups of species, without so much as a thought.

Whenever you mention extinctions, people usually think of dinosaurs, but the funny thing about dinosaurs is that they didn't actually go extinct. There was an event 65 million years ago, at a point in time that separates the Cretaceous period from the Paleogene period. That moment in geological time is called the Cretaceous-Paleogene boundary, or K-Pg boundary. It was then that *almost* all the dinosaurs disappeared, but a few did survive. In fact, their descendants live on to this day, and you see them all the time. Those descendants are the birds, and because all birds are direct descendants of the dinosaurs, birds, by definition, are also dinosaurs.* (Watch a chicken run and tell me it doesn't look like a velociraptor from *Jurassic Park*.)

Because the dinosaurs didn't go extinct, paleontologists are always careful to call the group that did die off at the K-Pg boundary the "non-avian dinosaurs." That's their way of saying "all the dinosaurs except the birds."

The K-Pg extinction may have wiped out charismatic animals like *Tyrannosaurs* and *Triceratops*, but the non-avian dinosaurs weren't the only ones that disappeared. Flying reptiles called pterosaurs did too (they weren't technically dinosaurs), as did the big swimming reptiles that looked like Loch Ness monsters (also not dinosaurs) and a whole bunch of other plants and animals. In all, roughly 70 to 75 percent of all species on Earth disappeared.[27]

The evidence is very strong that the K-Pg extinction event

* http://www.xkcd.com/1211/

began when a massive meteor smashed into what is now Mexico, right near Mérida, on the Yucatán Peninsula. How big was it? For context, think back to that fireball that lit up the sky over Chelyabinsk, Russia, in February 2013. That meteor was about fifty-five feet across[28]—big enough that when it burned up in the sky, it released a flash of light brighter than the sun and sent out a shock wave powerful enough to knock over building walls, shatter windows, and injure more than a thousand people. Millions of rocks of various sizes hurtle into our atmosphere each year, but the Chelyabinsk meteor is the biggest one Earth has seen in the past century.

The meteor that wiped out the non-avian dinosaurs? Oh, *slightly* bigger.

It left behind a crater about 125 miles in diameter, which means the asteroid itself was probably about 6 miles in diameter. Six miles! That makes it more than five hundred times wider than the Chelyabinsk meteor, and more than 200 million times heavier.[29]

Animals standing right under it would have been vaporized by the fireball when that impact occurred, and those nearby would have been killed by the wildfires and tidal waves that spread out from the impact site, but the global-scale problems came in the ensuing months. The dust kicked into the atmosphere by that impact would have blocked out sunlight, and since energy coming into the living world must be harnessed by plants from sunlight (as you'll recall from the chapter on gluttony), this debris cut off the flow of energy not only to plants but also to all animals. By the time the dust settled, months or even years later, most of Earth's species were dead.

Events like the asteroid at the K-Pg boundary are part of nature, as are the smaller-scale events we've seen in our lifetimes, like earthquakes, tsunamis, hurricanes, tornadoes, and blizzards. When natural disasters happen, animals die, and when the event is big enough, a whole species or multiple species can vanish. Those extinctions are an important factor in how life has evolved on our planet. For one thing, mammals were only able to become things like horses, bats, and apes because the non-avian dinosaurs were gone.[30] If that asteroid hadn't hit, if all that death hadn't happened, humans wouldn't exist. By the same token, humans will surely someday experience an extinction of our own. That's just how nature rolls.

The mass extinction at the K-Pg boundary wasn't the only one in Earth's history. In fact, it wasn't even the biggest. Before dinosaurs ever existed, an extinction event 251 million years ago killed 95 percent of all species on Earth. That mass extinction, between the Permian and Triassic periods, is (not surprisingly) called the Permo-Triassic boundary, or P-Tr boundary. Some people, however, just call it the Great Dying. Almost every single kind of plant, insect, fish, amphibian, and reptile disappeared, including some reptiles as big as bears. All that was left was a tiny sliver of the biodiversity that had existed before.

The P-Tr extinction seems to have been triggered by giant volcanoes in what is now Siberia. Volcanoes don't sound that threatening, but we're not talking about lava flowing down a hillside. We're talking about massive fields of volcanoes that burped up enough lava that it ultimately covered an area of 1.6 million square miles—roughly half the area of the contiguous United States. The lava would have been bad for the animals right at the volcanoes, but the gases that came out wreaked havoc worldwide. They triggered a cascade of environmental changes that animals on land

and in the oceans simply could not handle—poisonous gases, steep drops in aquatic oxygen levels, and runaway global warming with increases of around 11°F (6°C). To give you some idea of how bad it was, in some places the soil was completely wiped away, killing everything that had once lived there, except the molds and other fungi that could cling to the bare rock left behind. Despite the magnitude of that carnage, a few living things did manage to survive, and after a few million years, their descendants evolved to fill the roles of some of the creatures that had died. Some of those new creatures, for example, were the dinosaurs.[31]

The K-Pg and P-Tr extinctions are the best-known mass extinctions, because there are heaps of fossils from those events. But there's one other extinction that most experts believe was even more catastrophic. It's a little bit more mysterious than the others, though, because it happened so long ago that everything that lived was single-celled. And all the fossils of single-celled organisms look *very* similar, making it hard to tell exactly which species survived and which did not. But it's an extinction with special relevance for us, because unlike the K-Pg and P-Tr extinctions, which were caused by meteors and volcanoes that you can't really blame on anyone, that early extinction was caused by living things.

It happened 2.4 billion years ago, and it was devastating. Just about every form of life on Earth died, killed by a poisonous gas. That poisonous gas had only existed on Earth before in trace amounts, but suddenly it increased in concentration more than a thousandfold, wreaking death on land and in the seas. The gas was oxygen, and it came out of algae as a waste product from the new trick they'd just learned, photosynthesis. The name for that massive extinction is the Great Oxygenation Event.[32]

It might be hard to think of oxygen as a poisonous gas, but to many creatures it is. In fact, the word *poisonous* almost doesn't do it justice. Oxygen is horrible, and to live in a world of it, you need an arsenal of molecular defenses at the ready.

Oxygen loves electrons and will rip them off anything it can get near. Iron rusts because oxygen rips the electrons out of it. When you see a piece of wood burn, you're watching the oxygen all around it tear the electrons out of it, releasing energy as heat and light. (That's why fires need oxygen.) When our bodies break down sugars, we're using a complex molecular machinery to harness the energy that is released when the oxygen we breathe rips the electrons out of the sugars we've consumed.

Living things didn't have that kind of molecular machinery 2.4 billion years ago, so when oxygen invaded their bodies, they were corroded alive. The oxygen was coming out of early algae that were just doing their best to survive and as a result ended up pumping oxygen out of their bodies. This had the consequence of changing the rules of life for the other creatures of Earth. Most organisms had been doing just fine breathing gases like methane, but after the Great Oxygenation Event, either they found a way to cope with oxygen or they died.*

Today, oxygen makes up more than 20 percent of the atmosphere, and animals have adapted to use oxygen's volatility to their advantage. You can get almost twenty times more energy out of oxygen than you could ever possibly get from methane, so

* In those places on Earth that oxygen hasn't reached, like hydrothermal vents at the bottom of the ocean, inside geysers, or deep within some caves, many of those non-oxygen-breathing organisms still thrive. That's why you often hear people say that living things from those places might give insights into what life might be like on other planets where oxygen doesn't exist.

looking back you could try to argue that the Great Oxygenation Event was a good thing. It's not like Earth was getting things ready for us, though. Humans breathe oxygen because that's what we evolved with. We've played with the cards we were dealt.*

The wrath of nature goes well beyond the pain and suffering of a person bitten by a rattlesnake. Time and time again, living things are swept off the face of the Earth before new forms of life grow again. All three of the extinctions I mention serve as reminders of the wrath inherent to nature, but they're also good reminders that life carries on despite them. There's nothing permanent about Sam or his DNA or any living thing on the planet. DNA churns away, making meat robots, and those meat robots do their best to survive, subject to the whim of fate. Sometimes the rules change suddenly, but that's just how the cookie crumbles.

Like any other species before us that has enjoyed some time in the sun on this planet, our days here are numbered. There was nothing wrong with the methane-breathing cells that faced those first clouds of oxygen. They were just unlucky. The pterosaurs might well have survived through to today had it not been for a wayward asteroid. Humans are no different. A stroke of bad luck will one day put us out like a cigarette. In fact, that bad luck may well be our own brains, which have set to work causing nuclear meltdowns, wars, and rapid climate changes on a par with what has happened during other extinction events. We've wiped out many animals in the last century, including the Tasmanian wolf,

* There's something beautiful to me about the fact that when a yoga class takes a deep breath of that clean, rejuvenating oxygen, they're taking in a gas that wiped out almost every living thing on Earth. The world doesn't provide for living things; it's the living things that have evolved to thrive on whatever happens to be around.

the passenger pigeon, and the Caribbean monk seal. Perhaps you could say we are playing the role of the asteroid this time.

I find the fact that humans are causing animals around the world to face extinction overwhelmingly depressing, but I do get a sliver of solace knowing that no matter how badly we screw this world up, something new, and different, will thrive. Earth bounced back from those other disasters. We will never get the animals we're killing back, but in 10 million years, the world just might look like nothing ever happened.

It's fascinating how differently we humans treat the fate of our own DNA, depending on the time scale. We protect ourselves and our children at any cost, but when we think a few generations down the road, we're barely willing to be inconvenienced. It makes sense. Our DNA has programmed us to live in the present—to protect those of our progeny that are on the Earth at the same time we are. Our DNA can't think a century ahead to what climate change will do, or overfishing, or habitat loss. Besides, saving the environment would benefit other humans' DNA as much as our own. Because that wouldn't give our own DNA a selfish competitive advantage, it's not really part of our DNA's job description. Our instincts weren't built to handle the kind of power our species now wields over the planet. Nature's been playing rough with us for so long, maybe it's no surprise that we're being so rough with her now.

I certainly don't want Sam to have to raise his own kids in a world without black rhinos, but it looks like that's what's going to happen.* However, I have the luxury of worrying about that

* The number of black rhinos killed in 2012 was 668, almost 15 percent of the entire population. Compare that with 2007, when only 13 were poached, and you get a sense of how rapidly things have deteriorated for those animals in the last decade (Biggs et al. 2013).

because the odds are I'm not one of the parents who will lose a child to a venomous snake bite this year. I only get to think about long-term consequences because my DNA's short-term needs are met. If we're going to ask people around the world to preserve the plants and animals we think deserve to live here, we're going to have to help those humans escape the wrath of nature first.

7

PRIDE

Rise Up, Meat Robots

Ever since I started writing this book, I've been trying to reconcile the harshness of nature with the love I experience when I spend time with Sam. The examples I've uncovered along the way have made it even more clear to me than ever just how horrible we creatures of nature can be, but I've kept my eyes open for that one exception—just one nonhuman species with compassion beyond its own DNA's needs. If such an animal could be found, then maybe true fatherly love—something more than the DNA-driven self-interests of a dad—might exist. I looked for that exception everywhere, but writing this book, the closest I ever got was vampire bats.

As we discussed in the chapter on sloth, vampire bats share food with unrelated members of their own species. However, this can't be counted as selfless behavior because by participating in the food-sharing program, bats end up receiving food from other bats on the nights they fail to find food themselves. The fact that bats end up helping nonrelatives is, if anything, just an emergent phenomenon that happens when the bats act selfishly.

Clearly vampires weren't the selfless animals I was looking for, but if a selfless animal *does* exist, I figured Gerry Carter would know about it. Gerry, an undergrad during my PhD on vampire bats, helped me during fieldwork in Trinidad. He has since started a PhD of his own, focused on the food-sharing behavior of the vampire bats. He's very smart, and he's been working his way through experiments and literature surveys, all concerning the question of animal kindness. No one knows more about that stuff than Gerry does.

I called Gerry up and asked him if he knew of any animal that willingly performs a behavior that helps the DNA of other animals, even though it comes at a cost to its own DNA's survival. As I had suspected, he confirmed that he had never come across such an example. Sometimes an individual would sacrifice itself to help another individual, like the *Leptothorax* worker ants (from the chapter on envy) that sacrifice themselves to protect their queen. But cases like those all result in the DNA getting passed on. Gerry hadn't seen a single example of self-sacrifice where the costs to the animal's DNA outweighed the benefits.

To me, that was the nail in the coffin. Animals are selfish, so pure love, with no strings attached, can't exist out there. What I feel for Sam is just another case of an animal looking after its DNA.

I told him a bit about this book and what his answer meant for my book's conclusion, that my love for Sam was really nothing special. Sometimes science doesn't give you the answer you were hoping for.

Gerry laughed for a second, then without pausing, told me that he wholeheartedly disagreed.

"Just because you understand the mechanism doesn't make it imaginary. You just identified where it came from. How does that make it any less authentic? Have you ever heard Feynman talk about a flower?"

"Yeah," I said.

"Well, it's the same thing."

I knew the Feynman thing Gerry was talking about quite well. I'd tried to use it in conversations about the value of science many times. It comes from a slightly grainy filmed interview with the brilliant Nobel Prize–winning physicist Richard Feynman, who, smiling as he always seemed to be, talks about the aesthetics of science.* An artist friend, he says, has argued that a scientist can't appreciate a flower because he takes it all apart until it becomes dull. Feynman, in his disarming New York City accent, says,

> *I can appreciate the beauty of a flower. At the same time, I see much more . . . I could imagine the cells in there, the complicated actions inside, which also have a beauty. I mean it's not just beauty at this dimension, at one centimeter; there's also beauty at smaller dimensions, the inner structure, also the processes. The fact that the colors in the flower evolved in order to attract insects to pollinate it is interesting; it means that insects can see the color. It adds a question:*

* You can watch it on YouTube. It's only a minute and a half long: http://www.youtube.com/watch?v=ZbFM3rn4ldo.

Does this aesthetic sense also exist in the lower forms? Why is it aesthetic? All kinds of interesting questions [by] which the science knowledge only adds to the excitement, the mystery, and the awe of a flower. It only adds. I don't understand how it subtracts.

Feynman's view is the opposite of "Ignorance is bliss." To him, every new understanding you have about something makes it more mysterious and wonderful. To Feynman, the questions we ask about a flower are wonderful, but there's no reason to fear that we'll lose that wonder if the questions are answered, because every time you learn something new, you're rewarded with new, deeper mysteries that engage you even more. Scientists don't get bored after learning about the world. They just keep digging deeper. And the deeper they dig, the more wonderful the mysteries get.

I thought back to the first time I saw those vampire bats—how seeing their faces was enriched by all the scientific facts I'd read about them. So why, then, was I having the opposite reaction to information about fatherly love? Why was knowing about evolution making me think my love for Sam was *less* beautiful instead of *more* beautiful? Gerry's argument was to consider Feynman's ode to a flower as advice to a father. The fact that I understand the evolutionary origins of love should not have to "subtract" from its beauty. Knowing that my love for Sam comes from millions of years of evolution, if anything, should make that love *more* real.

One of the things I've loved most about my life in science has been the quality of the friends I've made. It's wonderful to spend time with smart people who challenge you and force you to question your own beliefs. I'd been plowing through scientific papers, weaving their stories together for this book, and been unable to resolve this crisis about Sam. But Gerry could instantly see the flawed assumption my crisis was based upon. Maybe he'd spent time thinking about this very problem for his research, or maybe he just understood my situation better because he's not a dad, so he wasn't caught up in it. Either way, I'd assumed that a father's love could not be pure and real if it had been built by the selfish process of biological evolution. By asking me to defend that assumption, Gerry made my whole argument crumble.

That conversation with Gerry was like flicking a switch in my brain. If a bear bites me with the teeth that were built by evolution, the pain I feel is real. By the same token, if I love my son with emotions that evolved, that love is real too.

When I had that botfly in my head, there was a thrill the whole time. I was participating in the bloodbath of nature. I was really experiencing what it meant to be a creature on this planet. I was doing battle with a parasite, and even though it kind of sucked, it was fun. I felt like I was part of nature. And yet when I took part in the oldest tradition that life forms have, the tradition of reproducing, I somehow talked myself into believing I wasn't part of nature anymore. I couldn't have been more wrong. Having Sam has given me a new way of connecting to the natural world.

\\\\\

There are seven deadly sins, and so far I've done my best to convince you that nature does a better job committing them than humans do. But I've only made it through six of the seven, and when it comes to pride, I think humans might actually take the trophy.

Pride is the idea that you're different from everyone else and that the normal rules shouldn't apply to you. This is where humans really shine. We once believed (and some people *still* believe) that Earth is at the center of the universe, that a God created us separately from the animals, and that humans have souls but animals do not. Each time scientists have uncovered evidence that has chipped away at one of those theories, society has taken decades or even centuries to accept the facts. We now know that we live on a rock that floats vulnerably in space, that we're built of molecules, and that our thoughts result from electrical impulses. If you want to feel important, those are tough pills to swallow, and humans haven't had a very easy time letting go of their pride to accept those ideas.

But even among those revelations, the theory of evolution seems to have been especially hard for humans to accept. With more than 150 years of scientific progress since Darwin wrote *On the Origin of Species*, the theory of evolution by means of natural selection still ruffles feathers. Natural selection is the only plausible explanation scientists have come up with to make sense of their observations of the natural world. In the century and a half since it was proposed, that theory has been tested over and over and over again, but no one has succeeded in proving it wrong. That's why scientists treat it as a fact.* But despite that overwhelming avalanche of evidence, there are still millions of otherwise well-educated people who refuse to accept that evolu-

* In 1973, Theodosius Dobzhansky wrote an essay in the *American Biology Teacher* with the eleven-word title "Nothing in Biology Makes Sense Except in the Light of Evolution." That sentence just nails it, and the essay does too. It's a classic, well-written piece, and I recommend it if you're looking for ways to argue evolution skeptics out of the dark (Dobzhansky 1973).

tion happens. We might be fine with the idea that the Earth goes around the sun, but society still hasn't completed its transition to acceptance of evolution.

Perhaps one reason that people tend to use the word *natural* in such a positive way is that it's a way of helping us through that transition. Saying that nature is wonderful lets us accept that we evolved from nature without letting go of the idea that we're special. Instead of saying we're just lowly animals, we've elevated the other living things to a spiritual plane on par with godliness. It's not such a blow to your ego to acknowledge that you evolved from nature if you tell yourself that nature is perfect.

The truth of the matter, though, is that the creatures of Earth are playing dirty with one another, fighting over the energy necessary to make DNA copies of themselves. They're not holy, benign creatures working together in harmony. It looks like that from a distance, just as New York looks clean from its skyline, but down at Earth's street level the creatures are locked in a high-stakes battle. It's a bloodbath, and we evolved in the midst of it. But however ruthless the nature we evolved in might be, I think human pride is *warranted.*

We *are* different from the other living creatures. The normal rules *don't* have to apply to us. Just because we evolved in nature doesn't mean we can't break new ground. In fact, human pride might be just the thing we *need* to save nature from us.

The mice of Gough Island are doomed because they're going to eat themselves out of house and home. Just think what would be possible if they could stop, realize the path they're on, and make adjustments as a group. They can't, though. They're mice. Even though it would benefit all of them to back off on the meat consumption and bring their populations under control, natural selection just can't get them there. Even though in the long term, all mice could win by taking that strategy, eating less or having fewer babies would cause conscientious mice to lose out to any mice that kept on being selfish jerks. Evolution, as a process, can't deal with problems like that. The mice won't be saved by their own instincts.

But we're not mice. Instead of pretending we are, let's use our massive brains to come up with some solutions. Let's stop assuming our natural instincts are the way to go and start acting intelligently. Let's have a little pride in humanity.

Earth's human population hovered around 1 billion for centuries, but our numbers have been rapidly climbing ever since the Industrial Revolution. Today there are 7 billion of us. Throughout our expansion, we've wiped out animals and plants everywhere we've gone, and that decimation continues to this day. Humans have wiped out big, charismatic animals every time we've arrived on a new island or continent, but that doesn't mean we need to do that to the whole planet.

Part of the problem is that we can live our lives with no accountability for the environmental consequences of our choices. The person who uses the remote starter to warm up their SUV for forty-five minutes before driving it to work will wake up in ten years on the same planet as the person who rode a bike to work all winter. People get their benefits as individuals, but we all pay the costs together, same as the doomed mice of Gough Island.*

The problem's not just fossil fuel emissions either. Humans are killing rhinos, elephants, gorillas, and countless other endangered animals for no reason other than to make a few bucks. Put

* This problem is called "the tragedy of the commons" (Hardin 1968).

another way, those of us in the Western world have allowed things to get so unfair that while I stop at a Starbucks drive-through on the way home to get a fifth of my daily required calories from a sugar-filled iced drink, someone on the other side of the world is eating an endangered fruit bat because they have no other way to get the protein. That's where acting "naturally" has gotten humans so far.

We have a choice. One option is to keep on focusing on the short term and let the planet fend for itself against our selfish individual wants and needs. It's not like the world would end. Sure, we won't have pandas or tigers or blue whales anymore, but evolution will continue, just as it has after other extinctions, and in a few million years, some new group will start to fill the roles in nature vacated by the animals we know today.

The alternative, though (am I hitting you hard enough over the head with this?), is for us to stop acting like mice and treat biodiversity as though it matters. We need to act as though nature has value beyond calories or dollars, even though our DNA doesn't have that appreciation built in. However deadly and selfish and brutal nature may be, it's unique, beautiful, surprising, and more valuable than words can express.

One important step in protecting nature is to make an effort to connect with it. The experience I got from having a botfly can't be mimicked by any human-built theme park or 3-D movie. When you see something real, for yourself, it makes a difference to the rest of your life. For me, seeing a real-life sac-winged bat changed me much more than reading about their salting behavior. You can watch all the documentaries about vampire bats you want to, but when you stick your head into their cave and let them scream at you, it's different.

Take a moment to think about your own experience with this. Have you ever seen a wild animal that made your jaw drop? A whale? A bear? An owl? A sea turtle? Have you ever gone snorkeling or hiking or even just watched birds at the feeder? Think about the value those experiences give you as a person. Why not make a decision to fill your life with more of those experiences? Visit parks. Spend your dollars on ecotourism in other countries. Take your friends and family members with you. Get out there and remind yourself why nature is worth preserving.

Another great way to help protect the world is to support science. Science is how we learn what species are out there and come up with plans to preserve them. By taking pride in our intellect and letting our species push the boundaries back, we're finding new ways to interact with nature, to protect our planet, and ultimately to protect ourselves. Some technologies, like genetic engineering, scare a lot of people because they seem so unnatural, but that might well be the best thing about them. It's only by looking beyond what's already been done that we can have our best hope of improving our trajectory. We have to have open minds about technology if we want to survive.

Every plant or animal we've ever changed through breeding has been genetically modified. Now that we understand the molecular basis of those changes, we can make them more quickly and effectively than ever before. There's good reason to be wary of large multinational corporations controlling the availability of foods, but there's no reason to fear genetic engineering as a technology. Just because it feels unnatural doesn't mean it's bad. We have to let go of that knee-jerk assumption.

And just as the tools of agriculture must change, so too will the scale. It might seem appealing to dream of a world where we get all our foods from nature, without big farms, but there are just too many people on Earth now for that to be a solution for everyone. With an expected population of 9 billion within the next forty years, we're going to need agriculture to be more efficient than ever before. I know big farms might not have the appeal of small, family-run ones, but big farms can be more efficient. So if you're going to cut down hectares and hectares of the Amazon rainforest to set up farms so people can eat, it's far better to grow as much food per unit area as possible so that we can cut down as little as possible. The dream that there's a natural way to live without agriculture just doesn't work in a world of 9 billion people—at least not for all of us.

Scientists are scrambling to find ways for our species to survive. They deserve all the support they can get.

\\\\\\

Moving away from this false ideal of "natural" living would be good for sustainability, but it's also important socially. What's natural about gender equality? What's natural about human rights? Nothing. A sea lion torn to shreds by a killer whale has no rights, and the female pintail duck doesn't get to file legal proceedings against the lecherous males who attacked her. Human rights are a fundamental component of the future we're trying to create, so let's call them what they are—unnatural and essential.

Rape has been part of the behavior of our species since before we were humans; we inherited it from our prehuman ancestors. Let's leave rape behind. There's no use for rape in the human world, and there's no place for people who think they can justify rape because it's natural. When a woman gives birth, let's not force her into making difficult decisions based on some imaginary "natural" way to do so. Let's give her the freedom to choose the experience she wants, secure in the comfort that modern medicine is there when she needs it. Let's also make that medicine available to people around the world, so that no woman has to experience pregnancy in fear and with unnecessary pain. There is no justification for sentencing eight hundred women to death that way each day, no matter how natural that risk might be. Let's let same-sex couples marry, not because male bats have sex with one another sometimes, but because it's the right thing to do.

Nature has no place in discussions of social justice. We're humans—we evolved in nature, but we can do better than the natural order. Let's take a little pride in the fact that we've invented human rights. It doesn't matter whether they're natural or not. We're animals, but we don't have to act like them.

What I'm really advocating for is a rebellion of the human meat robots against our DNA oppressors. We're hardwired to be selfish, to put our DNA's needs ahead of the other people and ecosystems around us. It's natural to care more about how hot your coffee was this morning than the fact that while you were drinking it a five-year-old girl in Tanzania died from malaria. But so long as people born in some parts of the world are doomed to hunger, poverty, and war, the progress enjoyed by the rest of us can't take effect. Biodiversity will slip away because we just weren't built to deal with issues of a global scale. But just because we weren't *built to* do that job doesn't mean we *can't* do it.

Find a nonprofit that is working on an issue you care about and join it. Consider the World Wildlife Fund, Amnesty International, or even Bat Conservation International. Read books. Vote for political parties that respect women's rights, environmental sustainability, social programs, and basic science research. Tell people why you vote that way. Change minds. Volunteer in your community. Volunteer in someone else's community. Work with kids. Do something to improve other people's lives.

Just rub it in your DNA's face.

It's quite possible that the last time organisms on Earth had as much influence on the globe as humans do today was during the Great Oxygenation Event. But our legacy doesn't need to be one of destruction. We have the power to choose our own destiny, so let's create a utopia here on Earth, based on human rights, equality, and environmental sustainability.

To do nothing about those issues and stay immersed in our own personal, selfish experiences is the most natural thing to do. So give your natural instincts the finger, and be unselfish. It's not enough to feel good when you see other people effecting those changes. For this to work, we need to do it as individuals. This is between you and the molecules inside you. Make the world a better place, even though it will be more work for you and cost you more money than doing nothing. We've been doing everything our DNA asked of us for long enough. Rise up, meat robots!

Shelby and I have big plans to take Sam around the world to see the wonders of nature. Someday maybe we'll even hunt out those bioluminescent ostracods, the emerald sea slug, or that vegetarian spider *Bagheera*. I can barely imagine the thrill of sharing those experiences with Sam. Maybe Shelby and I will even take Sam to our campground in Texas and scratch the sand together with him to call in those pallid bats. Until then, though, my plan is to just let myself be in love—in love with Sam, in love with Shelby, and in love with the fragile, wonderful natural world.

ACKNOWLEDGMENTS

First and foremost, thank you for reading this book.

There are many people to thank for helping me put this book together. I am grateful to Jeff Kleinman and Michelle Brower of Folio Literary Management for asking me what book I'd most like to write and then pushing me to think beyond the bat book I proposed in response. I'm also extremely grateful to Michelle Howry of Touchstone for believing in this book, for her edits and comments as I wrote, and for her patience and positivity throughout the writing process.

I also thank Susan Moldow, Stacy Creamer, David Falk, Sally Kim, Jessica Roth, Meredith Vilarello, Ana Paula de Lima, Martha Schwartz, George Turianski, Joy O'Meara, and Peg Haller, at Touchstone Books in New York, along with Kevin Hanson, David Millar, Alison Clarke, Sheila Haidon, Felicia Quon, Maximillian

Arambulo, Michelle Blackwell, Andrea Seto, and April Gibson at Simon & Schuster Canada, for their enthusiastic support of this book.

I am grateful to the scientists who have helped me shape my view of the world. Brock Fenton, John Hermanson, Sharon Swartz, and Tom Kunz supervised me through my graduate studies and postdocs, but there have been countless others—professors, postdocs, instructors, graduate students, and undergraduates who have also contributed to my education. I reached out to a few of those people while writing this book, and I thank Attila Bergou, Gerry Carter, Rulon Clark, Christina Davy, Yvonne Dzal, Leif Einarson, Brock Fenton, Patrick Flight, Megan Frederickson, Matt Heard, John Hermanson, John Hutchinson, Emily MacLeod, Troy Murphy, Matt Ogburn, John Ratcliffe, Shelby Riskin, Jaime Tanner, and Amity Wilczek for their helpful responses. I'm also grateful to the legion of scientists who wrote the two hundred plus scientific articles I drew upon while writing this book. None of this would be possible without their curiosity and hard work.

In the world of TV, Beth Hoppe has been very much like an academic advisor to me. She gave me my start on a show about evolution and then spent several years training me to be an effective communicator of science. The fact that I had the opportunity to write this book is a direct result of her support.

I wrote this book while working full-time at Discovery Canada's *Daily Planet*, and I thank my colleagues there for their patience while I did so. I'm especially grateful to Paul Lewis, Ken McDonald, Kelly McKeown, and Ziya Tong, but there are more than a hundred other people there who help make that show happen five days a week, and I thank them all for their support while

I worked on this book. Thanks also to Marilyn Haft for helping me juggle these various parts of my career.

I thank my mother, Mary W. Walters, who is a great editor (and also a fabulous author of fiction, by the way), for helping me tighten my writing and for supporting my wacky career in general. Mom, you're brilliant, and if I look half-smart sometimes, it's only because you teached me so good. I also thank my family for their support.

It's somewhat perverse that I took so much time away from Sam and Shelby to write a book about my love for Sam and Shelby, but somehow that's what happened. Shelby knew how much this book meant to me, helped me free up the time I needed to get it written, and offered many invaluable suggestions to help me improve the book itself. Thank you both.

Toronto
September 2013

NOTES

Introduction: Georgia on My Mind

1. For a detailed description of the human botfly's biology, complete with grotesque photos, visit the University of Florida Entomology and Nematology Department's "Featured Creatures" page: http://entnem dept.ufl.edu/creatures/misc/flies/human_bot_fly.htm.

Chapter 1. Greed

1. Lions do not kill zebras at random (Mills and Shenk 1992).
2. Biologists call this a distinction between *inter*specific competition (worrying about members of other species, like predators for example) and *intra*specific competition (worrying about members of your own species).
3. The selfish sheep study was by King et al. (2012).

4. Penguin huddle temperatures are from Gilbert et al. (2006).

5. This strategy is called asynchronous hatching. The snowy owl story is summarized by Murie (1929) and Parmelee (1992). Similar stories exist for blue-footed boobies (*Sula nebouxii*) and cattle egrets (*Bubulcus ibis*). There's a great review of "avian siblicide" by Mock et al. (1990).

6. The Verreaux's eagle (*Aquila verreauxii*) is also sometimes called a black eagle (Mock et al. 1990). The incident of 1,569 pecks was observed by Gargett (1978).

7. A thorough description of the sand tiger shark's "intrauterine cannibalism" can be found in Gilmore et al. (1983).

8. The whole story of the mice of Gough Island is told nicely by Cuthbert and Hilton (2004) and by Wanless et al. (2007).

9. The phrase "most important seabird island in the world" is used by a lot of people to describe Gough Island. I'm quoting Cuthbert and Hilton (2004).

10. The idea that mice can drive seabird extinctions is discussed by Wanless et al. (2007).

11. This event is called the Australian Pleistocene megafaunal extinction. For details about the species that disappeared, and a nice picture showing their sizes, see Flannery (1990).

12. The relative importance of hunting and wildfires to the Australian Pleistocene megafaunal extinction is discussed by Miller et al. (2005).

13. You guessed it: it's called the North American Pleistocene megafaunal extinction. To get an idea of the animals that lived in North America when humans first got there but that are now missing, see Janzen and Martin (1982).

14. The South Pacific extinction events happened in the Holocene, after the Pleistocene extinctions of Australia and North America, with two-thirds of extinctions happening between first human

settlement and European colonization of an island (Duncan et al. 2013).

15. The nautical disasters were analyzed by Elinder and Erixson (2012).

Chapter 2. Lust

1. The lives of *Antechinus* are reviewed by Naylor et al. (2008).
2. The researchers who chased the spiders around with paintbrushes are Li et al. (2012).
3. The Korean eunuch data are from Min et al. (2012).
4. These statistics about female survival of pregnancy come from the World Health Organization (2012).
5. Information about hyenas comes from Watts et al. (2009) and Glickman et al. (2006).
6. The social structure of spotted hyenas is nicely laid out by Watts et al. (2009).
7. You can watch Todd Akin make his political-career-ending statements at http://fox2now.com/2012/08/19/the-jaco-report-august-19-2012/.
8. Information about the mating habits of the northern pintail are from Sorenson and Derrickson (1994). Penis length data are from Brennan et al. (2007).
9. These comparisons across species have been made by Brennan et al. (2007).
10. This information about garter snakes comes from a paper by Shine et al. (2003).
11. The biology of bedbugs, from feeding to mating, is summarized by Reinhardt and Siva-Jothy (2007).
12. What it is specifically that female túngara frogs like to hear is described by Akre et al. (2011).

13. Unisexual reproduction among vertebrates is discussed by Neaves and Baumann (2011).

14. You can read all about penis fencing in a paper by Michiels and Newman (1998).

15. That estimate of 1 billion years is based on a paper by Butterfield (2000).

Chapter 3. Sloth

1. Global obesity data are from Swinburn et al. (2011).

2. Parasites influence how energy flows through ecosystems, how well the animals in ecosystems can compete, and, ultimately, how many different species an ecosystem can hold. Therefore, the more parasites you can find in an ecosystem, the healthier that ecosystem can be (Hudson et al. 2006).

3. Details about vampire bat saliva are available in Tellgren-Roth et al. (2009). The details about hunting come Greenhall and Schmidt (1988).

4. The common vampire bat jumps with a peak force equivalent to roughly 9.5 times its own body weight in around thirty milliseconds, sending its body upward at a speed of 7.8 feet per second (Schutt et al. 1997).

5. Biologists have been excited about blood sharing by vampire bats ever since it was first described by Wilkinson (1984). Gerry Carter has done some great follow-up work since then on the rules by which the bats decide for whom they will puke up food (Carter and Wilkinson 2013).

6. For details about the stretching stomach of the common vampire bat, complete with X-rays of bats that have fed on barium-laden blood, see Mitchell and Tigner (1970).

7. Squirrels have roundworms (Crompton 2001), birds have feather mites (Proctor 2003), giant pandas have at least six kinds of parasitic worms (Zhang et al. 2011), and emperor penguins are infected with tapeworms, lice, and the bacterial disease chlamydia (Barbosa and Palacios 2009).

8. Almost 50 percent of all *known* animal species are parasites, and Poulin and Morand (2000) make some great logical arguments about why the majority of the species we *don't* yet know are probably parasites. (In a nutshell, it's because they're harder to find.)

9. For more details about the lives of mosquitoes, see Christophers (1960).

10. The US Centers for Disease Control and Prevention (CDC) has a very informative website with data about parasite life cycles and their effects on humans. The page about malaria is at http://www.cdc.gov/malaria/.

11. The CDC webpage about elephantiasis is at http://www.cdc.gov/parasites/lymphaticfilariasis/.

12. The raccoon roundworm that can burrow through your body, eat your organs, and kill you is *Baylisascaris procyonis*. The eyeball-eating amoeba (which wasn't a human parasite at all until contact lenses were invented, by the way) is *Acanthamoeba keratitis*. The pinworm I'm talking about is *Enterobius vermicularis*, though there are several other kinds of pinworms you can get infected with as well.

13. Leeches pick an animal they want to feed on, then move toward it in bursts, continually getting updates about the animal's location (Harley et al. 2013).

14. These examples of facial ripping, including a very graphic image of a woman's face after a dog bit her cheek off, are from Koch et al. (2012).

15. Getting DNA of endangered animals out of leeches in their habitats is brilliant. Details of the procedure can be found in Schnell et al. (2012).

16. For more about bioluminescence by marine animals, and the relationships animals set up with light-producing bacteria, see Haddock et al. (2010).

17. To be honest, I don't know for sure that it was a Krøyer's deep sea

anglerfish that attacked them in that movie, but it was definitely an anglerfish of some kind.

18. This business of tiny males acting like parasites of females isn't restricted to the Krøyer's deep sea anglerfish. It's found in a bunch of different anglerfishes (Herring 2007).

19. The argument that male anglerfish don't really count as parasites, despite their lazy lifestyles, is made nicely by Vollrath (1998).

20. The CDC webpage for schistosomiasis is at http://www.cdc.gov/parasites/schistosomiasis/.

21. Whether the schistosome sets up shop in the blood vessels near the bladder or the intestines depends on the species. *Schistosoma haematobium* eggs come out in your urine, but the eggs of *S. mansoni* and *S. japonicum* exit in your feces.

22. For more about swimmer's itch, see Verbrugge et al. (2004).

23. The "dog on a leash" analogy comes straight from the scientific paper that describes all this (Gal and Libersat 2008).

24. The paper about the larvae spitting antimicrobial chemicals all over the place is by Herzner et al. (2013).

25. These numbers, about how common parasitoid species are, come from Eggleton and Belshaw (1992) and Feener and Brown (1997).

26. The biomechanics of poop launching by the skipper caterpillar are described in mouthwatering detail by Caveney et al. (1998).

27. Here's how I "put that in human terms": The insect is around 2 inches long and launches its poop 30 inches. That's 15 body lengths. If you multiply the body length of a 5-foot-tall woman 15 times, you get 75 feet.

28. Carl Zimmer has done a wonderful job of exploring these themes in his book *Parasite Rex* (2000).

29. Details about the effects of the *Toxoplasma* parasite on the behavior of

rats can be found in some outstanding papers from Robert Sopolsky's lab at Stanford (Vyas et al. 2007; House et al. 2011).

30. The estimates of the number of people infected with the *Toxoplasma* parasite come from Havlíček et al. (2001) and Montoya and Liesenfeld (2004).

31. The later during pregnancy the infection of the mother happens, the more dangerous it is for the baby. The threats posed by infection with the *Toxoplasma* parasite during pregnancy are reviewed by Wong and Remington (1994).

32. The reaction time data come from Havlíček et al. (2001); the traffic accidents data come from Flegr et al. (2002). The overall effects of *Toxoplasma* on humans are reviewed by Flegr (2013).

33. The potential influence of *Toxoplasma* on human cultures was explored by Lafferty (2006).

34. The ears of moths evolved as a direct response to echolocation by bats and work only for hearing the high-pitched calls of bats (Windmill et al. 2007). As a counterstrategy, one bat, called the spotted bat, has lowered its voice to frequencies that moths can't hear, but as a result the bat has become audible to humans (Fullard and Dawson 1997).

35. This whole mites-on-the-ears-of-moths system is described by Treat (1957).

36. The link between the mother's immune system and the sexual preferences of her sons was studied by Blanchard (2001).

Chapter 4. Gluttony

1. There's a nice discussion about why some plants and animals were domesticated while others were not by Diamond (2002). If you've ever wondered why you can't farm zebras and water buffalo, when their close cousins, horses and cows, are so easy to get along with, I recommend giving that paper a read.

2. Dr. Shah's case summary from the 2003 study is available for download at http://www.sudhirneuro.org/files/mataji_case_study.pdf.

3. The PDF of Dr. Shah's hypothesis, explaining how a human might be able to live without food, is available for download at http://www.sudhirneuro.org/files/fast_the_hypothesis.pdf.

4. For nonobese adults, the body is roughly 60 percent water (Ellis 2000).

5. Astronaut water consumption was modeled by Hager et al. (2010).

6. That's 30 lb. of batteries! Here's where I get that number: A person's resting metabolic rate is roughly 1,250 kcal/day (Tranah et al. 2011), which in different units is around 1,450 watt-hours. An alkaline AA battery gives you about 2.5 watt-hours, so 580 batteries would do the trick.

7. All oxygen-producing photosynthetic organisms come from a common single-celled ancestor that lived around then (Falkowski 2011).

8. This acacia-ant relationship is described by González-Teuber et al. (2012).

9. Elephants run into another species of acacia, which also uses ants, called the whistling-thorn tree. Elephants will eat those trees if there are no ants in them but avoid the trees when ants are present (Goheen and Palmer 2010).

10. The number 200,000 comes from Mithöfer and Boland (2012).

11. Cyanide lethality data come from Way (1984).

12. That number comes from Vetter (2000).

13. The "bombs" in this analogy are molecules of acetone cyanohydrin. The "detonator" is an enzyme called hydroxynitrile lyase. Bombs are kept in vacuoles and detonators in the tissues. When a vacuole gets broken open by a chewing animal, the enzyme breaks apart the acetone cyanohydrin to produce acetone and hydrogen cyanide (Vetter 2000).

14. One study of gorillas found that they ate eighty-four different kinds of plants, only two of which contained hydrogen cyanide (Rothman et al. 2006).

15. In Africa and Latin America, cassava is one of the three most important sources of food energy (Vetter 2000).

16. The story of zonal geraniums, quisqualic acid, and Japanese beetles comes from Ranger et al. (2011).

17. The gene pathway is called the Sonic Hedgehog signaling pathway. If you think that's a funny name, you're right. The pathway was named after a 1990s video game for Sega Genesis called Sonic the Hedgehog.

18. It's apparently the difference between the experience of being chewed versus having your juices sucked out that allows the plant to tell if a caterpillar or spider mite is at work (Leitner et al. 2005).

19. Karban et al. (2004) even showed that some plants can sense the alarm chemicals secreted by plants of other species.

20. All these questions about why plants release these chemicals are addressed in a nice review article by Heil and Karban (2010).

21. The recruitment of parasitic wasps against caterpillars by plants is described by Paré and Tumlinson (1999).

22. The emerald sea slug (*Elysia chlorotica*) feeds on an alga called *Vaucheria litorea* (Rumpho et al. 2008).

23. The aphid is called *Acyrthosiphon pisum* (Valmalette et al. 2012).

24. This bat's tongue really is spectacular to see (Muchhala and Thomson 2009; Muchhala 2006).

25. Improved seed germination after digestion has been shown for seeds that have gone through birds, monkeys, and bats (Fleming and Heithaus 1981).

26. Bananas come from Southeast Asia, and their domestication has taken humans thousands of years (De Langhe et al. 2009).

27. The habit of Sumatran orangutans (*Pongo pygmaeus*) eating slow lorises (*Nycticebus coucang*) is documented by Utami and Van Hooff (1997).

28. All this wolverine information comes from a review by Pasitschniak-Arts and Larivière (1995).

29. The 6 percent for African elephants, which average a weight of 1,700 kg., comes from Laws (1970). That spectacular 384 percent number comes from a 3.35-gram shrew eating earthworms (Morrison et al. 1957).

30. Here's the math: You can fit about half a million 3.35-gram shrews inside a 1,700-kg. (3,747 lb.) elephant. The shrew's consumption (384 percent body weight per day) is sixty-four times higher than the elephant's consumption (6 percent body weight per day).

31. The whale estimate of 165,000 kg. (363,000 lb.) comes from Lockyer (1976).

32. That estimate of 1,120 kg. (almost 2,500 lb.) of krill per whale per day comes from Goldbogen et al. (2011). Krill weigh around 2 grams each.

33. This whole process is explained by Vass (2001).

34. I've already acknowledged that some of the energy in ecosystems comes from hydrothermal vents, but that's in places like the deepest parts of the ocean or in pools of water at Yellowstone National Park. For the majority of the natural places humans visit, energy comes from the sun.

35. For a list of 203 crop plants humans have domesticated, see Meyer et al. (2012). It's kind of a fun list to look at.

36. Had Sam been born in the USA in 1911, his life expectancy would be 49.9 years (Shrestha 2006). According to the US Census Bureau website (http://www.census.gov/population/international/), a boy born in 2011 has a life expectancy of 78 years if he's born in the USA, or 81 years if he's born in Canada (which Sam was).

37. That estimate of 9 billion by 2050 comes from a meeting of the Food and Agriculture Organization of the United Nations (FAO 2009).

Chapter 5. Envy

1. The invasion of *Leptothorax* colonies by *Epimyrma* ants is described by Buschinger (1989) and Iyengar (2008).
2. That estimate of two hundred thieving ant species and the description of the "slave raids" by *Polyergus* ants come from a great article for popular audiences in *Scientific American* magazine (Topoff 1999). In case you're wondering, roughly 10,000 kinds of ants have been described so far, but entomologists think that might be about half of what's really out there.
3. The use of web vibrations by *Argyrodes* to observe the movements of orb-weaving spiders was studied by Vollrath (1979).
4. The use of stolen food as a nuptial gift by *Argyrodes* spiders is described by Uetz et al. (2010).
5. The relationship of the tiny *Curimagua* spider and the much larger *Diplura* spider is described by Vollrath (1978).
6. I would guess that there are other spider species like *Curimagua* hiding out there that we simply don't know about yet because they're so tiny.
7. The biology of the (mostly) vegetarian spider *Bagheera kiplingi* is described by Meehan et al. (2009).
8. Theft among lions, hyenas, and the other large African predators are documented by Iyengar (2008).
9. The fact that cheetahs will stop hunting if they so much as hear lions and hyenas nearby was documented by Durant (2000).
10. The avoidance of parks by African wild dogs to avoid lions and hyenas was documented by van der Meer et al. (2011).

11. Carrier (1984) was the first person to put forward the idea that humans are "born to run." That theory is fairly widely accepted now. See Bramble and Lieberman (2004) for a nice review.

12. The study of Hadza scavenging was done by O'Connell et al. (1988).

13. The capuchin experiments I describe here were done by van Wolkenten et al. (2007).

14. These dog experiments were done by Range et al. (2009), and you're welcome to try your own if you have access to two dogs.

15. The species for which fairness ("envy") experiments have been conducted are reviewed by Brosnan and de Wall (2012).

16. The sexual behavior of the sac-winged bat *Saccopteryx bilineata* is described by Voigt and von Helversen (1999) and by Voigt et al. (2005; 2008).

17. This information about toads comes from a classic paper by Sullivan (1983).

18. The frog in question, the European common frog *Rana temporaria*, is found across Europe, as its name suggests. The behavior of "clutch piracy" was seen in Spain (Vieites et al. 2004).

19. The mating habits of the frog *Rhinella* are described by Izzo et al. (2012).

20. The mating behavior of marine iguanas is described by Wikelski and Bäurle (1996).

21. The bioluminescent behavior of ostracods is described by Rivers and Morin (2009).

22. The tendency of sneaker male cuttlefish to make themselves look like females is described by Hanlon et al. (2005).

23. This outrageous tale of amphibian sexual mimicry is told by Arnold (1976).

Chapter 6. Wrath

1. The story of Tilikum the killer whale is the subject of a great article in *Outside* magazine by Zimmermann (2010) and of an independent film titled *Blackfish* (2013).
2. Information about killer whale food comes from Ford et al. (2011).
3. A description of killer whale behavior while hunting a really big whale can be found in Silber et al. (1990).
4. The impaling behavior of shrikes is described by Yosef and Pinshow (2005).
5. The accelerations of jellyfish stingers (nematocysts) were measured by Holstein and Tardent (1984).
6. The mechanism of action of box jellyfish stings is reviewed by Yanagihara and Shohet (2012).
7. The ability of blue dragons to eat harpoons without firing them was described by Thompson and Bennett (1969).
8. The blue dragon is just one of a number of animals that can eat and then use jelly harpoons. How it all works is reviewed by Greenwood (2009).
9. Cone snails produce analgesic compounds (Nelson 2004).
10. The estimate of the number of human deaths due to cone snail envenomation comes from Nelson (2004).
11. The mechanisms by which spider neurotoxins function are reviewed by Escoubas et al. (2000).
12. For a review of how the Australian funnel-web spider's venom works, see Nicholson et al. (2006).
13. The lethality of Australian funnel-web spider bites is reviewed by Isbister et al. (2005).
14. The giant fossil *Brontoscorpio* is described by Kjellesvig-Waering (1972).

15. All these numbers come from a short but fascinating review of scorpion biology by Polis (1990).

16. The biology of the pallid bat, *Antrozous pallidus*, is reviewed by Hermanson and O'Shea (1983). (Hermanson was my PhD advisor.)

17. The information about how pallid bats deal with scorpions comes from conversations I had with Dave Johnston, who did his PhD on pallid bats and has studied them extensively since.

18. The first version of the Schmidt Pain Index, from 1983, had much of the poetic writing that I like so much, but the list from 1990 has the most species on it (Schmidt et al. 1983; Schmidt 1990).

19. The experience that man had with the sting of a *Dinoponera* ant is described by Haddad et al. (2005).

20. The number of snake species known to science is tabulated at a website called the reptile database, http://www.reptile-database.org/db-info/SpeciesStat.html. When I last checked, the number of snakes (suborder Serpentes) was 3,432.

21. Estimates of numbers of deaths due to snake bites are from Warrell (2010).

22. A truly heartbreaking photograph of a girl's leg in exactly this condition, after she was bitten by a spitting cobra, can be found in Warrell (2010).

23. For the different ways snake venom can affect the blood, see Braud et al. (2000).

24. The difference in prognosis for a person bitten by a snake with one or the other of these neurotoxins (presynaptic or postsynaptic) is summarized by Del Brutto and Del Brutto (2012).

25. The prognosis for a victim of a Malayan krait is outlined by Warrell (2010).

26. The fact that rattlesnake venom helps the snake digest its food was shown by Thomas and Pough (1979).

27. Estimates I use in this chapter for the percentages of species that went extinct during extinction events come from Jablonski (1994).

28. That fifty-five-foot diameter estimate (seventeen meters) comes from Showstack (2013).

29. Estimates of the K-Pg meteor's size come from Urrutia-Fucugauchi et al. (2011).

30. The importance of the K-Pg extinction event for mammalian success has been quantified recently by Meredith et al. (2011).

31. Details about the P-Tr extinction and what caused it are reviewed by Benton and Twitchett (2003). The kinds of life that sprang up after that event are discussed by Sidor et al. (2013).

32. Details of the Great Oxygenation Event can be found in Sessions et al. (2009).

REFERENCES

Akre, K. L., H. E. Farris, A. M. Lea, R. A. Page, and M. J. Ryan. 2011. Signal perception in frogs and the evolution of mating signals. *Science* 333: 751–52. doi:10.1126/science.1205623.

Ancel, A., H. Visser, Y. Handrich, D. Masman, and Y. Le Maho. 1997. Energy saving in huddling penguins. *Nature* 385: 304–305. doi:10.1038/385304a0.

Anderson, J. T., T. Nuttle, J. S. Saldaña Rojas, T. H. Pendergast, and A. S. Flecker. 2011. Extremely long-distance seed dispersal by an overfished Amazonian frugivore. *Proceedings of the Royal Society of London B* 278: 3329–35. doi:10.1098/rspb.2011.0155.

Arnold, S. J. 1976. Sexual behavior, sexual interference and sexual defense in the salamanders *Ambystoma maculatum*, *Ambystoma tigrinum* and *Plethodon jordani*. *Zeitschrift für Tierpsychologie* 42: 247–300. doi:10.1111/j.1439-0310.1976.tb00970.x.

Baird, R. W., and L. M. Dill. 1996. Ecological and social determinants of group size in transient killer whales. *Behavioral Ecology* 7: 408–16. doi:10.1093/beheco/7.4.408.

References

Barbosa, A., and M. J. Palacios. 2009. Health of Antarctic birds: A review of their parasites, pathogens, and diseases. *Polar Biology* 32: 1095–115. doi:10.1007/s00300-009-0640-3.

Barrett-Lennard, L. G., C. O. Matkin, J. W. Durban, E. L. Saulitis, and D. Ellifrit. 2011. Predation on gray whales and prolonged feeding on submerged carcasses by transient killer whales at Unimak Island, Alaska. *Marine Ecology Progress Series* 421: 229–41. doi:10.3354/meps08906.

Benton, M. J., and R. J. Twitchett. 2003. How to kill (almost) all life: the end-Permian extinction event. *Trends in Ecology & Evolution* 18: 358–65. doi:10.1016/S0169-5347(03)00093-4.

Biggs, D., F. Courchamp, R. Martin, and H. P. Possingham. 2013. Legal trade of Africa's rhino horns. *Science* 339: 1038–39. doi:10.1126/science.1229998.

Blanchard, R. 2001. Fraternal birth order and the maternal immune hypothesis of male homosexuality. *Hormones and Behavior* 40: 105–14. doi:10.1006/hbeh.2001.1681.

Borgia, G. 1985. Bower quality, number of decorations and mating success of male satin bowerbirds (*Ptilonorhynchus violaceus*): An experimental analysis. *Animal Behaviour* 33: 266–71.

Botelho, J. B., and V. A. C. M. Weigel. 2011. The Satereé-Maweé community of Y'Apyrehyt: Ritual and health on the urban outskirts of Manaus. *História, Ciências, Saúde-Manguinhos* 18: 723–44. doi:10.1590/S0104-59702011000300007.

Bramble, D. M., and D. Lieberman. 2004. Endurance running and the evolution of *Homo*. *Nature* 432: 345–52. doi:10.1038/nature03052.

Braud, S., C. Bon, and A. Wisner. 2000. Snake venom proteins acting on hemostasis. *Biochimie* 82: 851–59. doi:10.1016/S0300-9084(00)01178-0.

Breene, R. G., and M. H. Sweet. 1985. Evidence of insemination of multiple females by the male black widow spider, *Latrodectus mactans* (Araneae: Theridiidae). *Journal of Arachnology* 13: 331–35.

Brennan, P. L. R., C. J. Clark, and R. O. Prum. 2010. Explosive eversion and functional morphology of the duck penis supports sexual conflict in waterfowl genitalia. *Proceedings of the Royal Society B* 277: 1309–14. doi:10.1098/rspb.2009.2139.

References

Brennan, P. L. R., R. O. Prum, K. G. McCracken, M. D. Sorenson, R. E. Wilson, and T. R. Birkhead. 2007. Coevolution of male and female genital morphology in waterfowl. *PLoS One* 2: e418. doi:10.1371/journal.pone.0000418.

Brosnan, S. F., and F. B. M. de Waal. 2012. Fairness in animals: Where to from here? *Social Justice Research* 25: 1–16. doi:10.1007/s11211-012-0165-8.

Buschinger, A. 1989. Evolution, speciation, and inbreeding in the parasitic ant genus *Epimyrma* (Hymenoptera, Formicidae). *Journal of Evolutionary Biology* 2: 265–83. doi:10.1046/j.1420-9101.1989.2040265.x.

Butterfield, N. J. 2000. *Bangiomorpha pubescens* n. gen., n. sp.: implications for the evolution of sex, multicellularity, and the Mesoproterozoic/Neoproterozoic radiation of eukaryotes. *Paleobiology* 26: 386–404. doi:10.1666/0094-8373(2000)026<0386:BPNGNS>2.0.CO;2.

Carrier, D. R. 1984. The energetic paradox of human running and hominid evolution. *Current Anthropology* 25: 483–89.

Carter, G. G., and G. S. Wilkinson. 2013. Food sharing in vampire bats: Reciprocal help predicts donations more than relatedness or harassment. *Proceedings of the Royal Society B* 280: 20122573. doi:10.1098/rspb.2012.2573.

Caveney, S., H. McLean, and D. Surry. 1998. Faecal firing in a skipper caterpillar is pressure-driven. *Journal of Experimental Biology* 201: 121–33.

Christophers, R. 1960. Aëdes aegypti *(L.), the yellow fever mosquito: Its life history, bionomics, and structure*. Cambridge, UK: Cambridge University Press.

Clark, R. W. 2005. The social lives of rattlesnakes. *Natural History* 114: 36–42.

Connor, R. C., R. A. Smolker, and A. F. Richards. 1992. Two levels of alliance formation among male bottlenose dolphins (*Tursiops* sp.). *Proceedings of the National Academy of Sciences USA* 89: 987–90. doi:10.1073/pnas.89.3.987.

Crompton, D. W. T. 2001. *Ascaris* and ascariasis. *Advances in Parasitology* 48: 285–375. doi:10.1016/S0065-308X(01)48008-0.

Cross, F. R., and R. R. Jackson. 2011. Olfaction-based anthropophily in a mosquito-specialist predator. *Biology Letters* 7: 510–12. doi:10.1098/rsbl.2010.1233.

Cuthbert, R. C., and G. Hilton. 2004. Introduced house mice *Mus musculus*: A significant predator of threatened and endemic birds on Gough Island,

South Atlantic Ocean? *Biological Conservation* 117: 483–89. doi:10.1016/j.biocon.2003.08.007.

Dawkins, R. 1976. *The Selfish Gene*. Oxford: Oxford University Press.

De Langhe, E., L. Vrydaghs, P. de Maret, X. Perrier, and T. Denham. 2009. Why bananas matter: An introduction to the history of banana domestication. *Ethnobotany Research and Applications* 7: 165–77.

Del Brutto, O. H., and V. J. Del Brutto. 2012. Neurological complications of venomous snake bites: A review. *Acta Neurologica Scandinavica* 125: 363–72. doi:10.1111/j.1600-0404.2011.01593.x.

Diamond, J. 2002. Evolution, consequences, and future of plant and animal domestication. *Nature* 418: 700–707. doi:10.1038/nature01019.

Dobzhansky, T. 1973. Nothing in biology makes sense except in the light of evolution. *American Biology Teacher* 35: 125–29. doi:10.2307/4444260.

Duncan, R. P., A. G. Boyer, and T. M. Blackburn. 2013. Magnitude and variation in prehistoric bird extinctions in the Pacific. *Proceedings of the National Academy of Sciences USA* 110: 6436–41. doi:10.1073/pnas.1216511110.

Durant, S. M. 2000. Living with the enemy: Avoidance of hyenas and lions by cheetahs in the Serengeti. *Behavioral Ecology* 11: 624–32. doi:10.1093/beheco/11.6.624.

Eggleton, P., and R. Belshaw. 1992. Insect parasitoids: an evolutionary overview. *Philosophical Transactions of the Royal Society of London B* 337: 1–20. doi:10.1098/rstb.1992.0079.

Elinder, M., and O. Erixson. 2012. Gender, social norms, and survival in maritime disasters. *Proceedings of the National Academy of Sciences USA* 109: 1322–24. doi:10.1073/pnas.1207156109.

Ellershaw, J. E., J. M. Sutcliffe, and C. M. Saunders. 1995. Dehydration and the dying patient. *Journal of Pain and Symptom Management* 10: 192–97. doi:10.1016/0885-3924(94)00123-3.

Ellis, K. J. 2000. Human body composition: In vivo methods. *Physiological Reviews* 80: 649–80.

Escoubas, P., S. Diochot, and G. Corzo. 2000. Structure and pharmacology of spider venom neurotoxins. *Biochimie* 82: 893–907. doi:10.1016/S0300-9084(00)01166-4.

Faith, J. T. 2011. Late Pleistocene climate change, nutrient cycling, and the megafaunal extinctions in North America. *Quaternary Science Reviews* 30: 1675–80. doi:10.1016/j.quascirev.2011.03.011.

Falkowski, P. G. 2011. The biological and geological contingencies for the rise of oxygen on Earth. *Photosynthesis Research* 107: 7–10. doi:10.1007/s11120-010-9602-4.

FAO. 2009. *How to Feed the World in 2050*. Rome, Italy: Report. Food and Agriculture Organization of the United Nations (FAO).

Feener, D. H. Jr., and B. V. Brown. 1997. Diptera as parasitoids. *Annual Review of Entomology* 42: 73–97. doi:10.1146/annurev.ento.42.1.73.

Fenton, M. B. 1983. *Just Bats*. Toronto: University of Toronto Press.

Fenton, M. B., E. Bernard, S. Bouchard, L. Hollis, D. S. Johnston, C. L. Lausen, J. M. Ratcliffe, D. K. Riskin, J. R. Taylor, and J. Zigouris. 2001. The bat fauna of Lamanai, Belize: Roosts and trophic roles. *Journal of Tropical Ecology* 17: 511–24. doi:10.1017/S0266467401001389.

Fenton, M. B., M. J. Vonhof, S. Bouchard, S. A. Gill, D. S. Johnston, F. A. Reid, D. K. Riskin, K. L. Standing, J. R. Taylor, and R. Wagner. 2000. Roosts used by *Sturnira lilium* (Chiroptera: Phyllostomidae) in Belize. *Biotropica* 32: 729–33. doi:10.1646/0006-3606(2000)032[0729:RUBSLC]2.0.CO;2.

Fittkau, E. J., and H. Klinge. 1973. On biomass and trophic structure of the central Amazonian rain forest ecosystem. *Biotropica* 5: 2–14.

Flannery, T. F. 1990. Pleistocene faunal loss: implications of the aftershock for Australia's past and future. *Archaeology in Oceania* 25: 45–55.

Flegr, J. 2013. Influence of latent *Toxoplasma* infection on human personality, physiology and morphology: Pros and cons of the *Toxoplasma*–human model in studying the manipulation hypothesis. *Journal of Experimental Biology* 216: 127–33. doi:10.1242/jeb.073635.

Flegr, J., J. Havlíček, P. Kodym, M. Malý, and Z. Smahel. 2002. Increased risk of traffic accidents in subjects with latent toxoplasmosis: a retrospective case-control study. *BMC Infectious Diseases* 2: 11. doi:10.1186/1471-2334-2-11.

Fleming, T. H., and E. R. Heithaus. 1981. Frugivorous bats, seed shadows, and the structure of tropical forests. *Biotropica* 13: 45–53.

Ford, J. K. B., G. M. Ellis, C. O. Matkin, M. H. Wetklo, L. G. Barrett-Lennard, and R. E. Withler. 2011. Shark predation and tooth wear in a population of northeastern Pacific killer whales. *Aquatic Biology* 11: 213–24. doi:10.3354/ab00307.

Fullard, J. H., and J. W. Dawson. 1997. The echolocation calls of the spotted bat *Euderma maculatum* are relatively inaudible to moths. *Journal of Experimental Biology* 200: 129–37.

Gal, R., and F. Libersat. 2008. A parasitoid wasp manipulates the drive for walking of its cockroach prey. *Current Biology* 18: 877–82. doi:10.1016/j.cub.2008.04.076.

Gargett, V. 1978. Sibling aggression in the black eagle in the Matopos, Rhodesia. *Ostrich* 49: 57–63. doi:10.1080/00306525.1978.9632631.

Garland, T. Jr. 1983. The relation between maximal running speed and body mass in terrestrial mammals. *Journal of Zoology London* 199: 157–70. doi:10.1111/j.1469-7998.1983.tb02087.x.

Gilbert, C., G. Robertson, Y. Le Maho, Y. Naito, and A. Ancel. 2006. Huddling behavior in emperor penguins: Dynamics of huddling. *Physiology and Behavior* 88: 479–88. doi:10.1016/j.physbeh.2006.04.024.

Gilmore, R. G., J. W. Dodrill, and P. A. Linley. 1983. Reproduction and embryonic development of the sand tiger shark, *Odontaspis taurus* (Rafinesque). *Fishery Bulletin* 81: 201–25.

Glickman, S. E., G. R. Cunha, C. M. Drea, A. J. Conley, and N. J. Place. 2006. Mammalian sexual differentiation: Lessons from the spotted hyena. *Trends in Endocrinology & Metabolism* 17: 349–56. doi: 10.1016/j.tem.2006.09.005.

Goheen, J. R., and T. M. Palmer. 2010. Defensive plant-ants stabilize megaherbivore-driven landscape change in an African savanna. *Current Biology* 20: 1–5. doi:10.1016/j.cub.2010.08.015.

Goldbogen, J. A., J. Calambokidis, E. Oleson, J. Potvin, N. D. Pyenson, G. Schorr, and R. E. Shadwick. 2011. Mechanics, hydrodynamics and energetics of blue whale lunge feeding: Efficiency dependence on krill density. *Journal of Experimental Biology* 214: 131–46. doi:10.1242/jeb.048157.

González-Teuber, M., J. C. Silva Bueno, M. Heil, and W. Boland. 2012. Increased host investment in extrafloral nectar (EFN) improves the efficiency of a mutualistic defensive service. *PLoS One* 7: e46598. doi:10.1371/journal.pone.0046598.

Greenhall, A. M., and U. Schmidt. 1988. *Natural History of Vampire Bats*. Boca Raton, FL: CRC Press.

Greenwood, P. G. 2009. Acquisition and use of nematocysts by cnidarian predators. *Toxicon* 54: 1065–70. doi:10.1016/j.toxicon.2009.02.029.

Haddad, V. Jr., J. L. C. Cardoso, and R. H. P. Moraes. 2005. Description of an injury in a human caused by a false tocandira (*Dinoponera gigantea*, Perty, 1833) with a revision on folkloric, pharmacological, and clinical aspects of the giant ants of the genera *Paraponera* and *Dinoponera* (sub-family Ponerinae). *Revista do Instituto de Medicina Tropical de São Paulo* 47: 235–38. doi:10.1590/S0036-46652005000400012.

Haddock, S. H. D., M. A. Moline, and J. F. Case. 2010. Bioluminescence in the sea. *Annual Review of Marine Science* 2: 443–93. doi:10.1146/annurev-marine-120308-081028.

Hager, P., M. Czupalla, and U. Walter. 2010. A dynamic human water and electrolyte balance model for verification and optimization of life support systems in space flight applications. *Acta Astronautica* 67: 1003–24. doi:10.1016/j.actaastro.2010.06.001.

Hanlon, R. T., M. Naud, P. W. Shaw, and J. H. Navenhand. 2005. Transient sexual mimicry leads to fertilization. *Nature* 433: 212. doi:10.1038/433212a.

Hansen, L. S., S. F. González, S. Toft, and T. Bilde. 2008. Thanatosis as an adaptive male mating strategy in the nuptial gift-giving spider *Pisaura mirabilis*. *Behavioral Ecology* 19: 546–51. doi:10.1093/beheco/arm165.

Hardin, G. 1968. The tragedy of the commons. *Science* 162: 1243–48. doi:10.1126/science.162.3859.1243.

Harley, C. M., M. Rossi, J. Cienfuegos, and D. Wagenaar. 2013. Discontinuous locomotion and prey sensing in the leech. *Journal of Experimental Biology* 216: 1890–97. doi:10.1242/jeb.075911.

Havlíček, J., Z. Gašová, A. P. Smith, K. Zvára, and J. Flegr. 2001. Decrease of psychomotor performance in subjects with latent "asymptomatic" toxoplasmosis. *Parasitology* 122: 515–20. doi:10.1017/S0031182001007624.

Heil, M., and R. Karban. 2010. Explaining evolution of plant communication by airborne signals. *Trends in Ecology and Evolution* 25: 137–44. doi:10.1016/j.tree.2009.09.010.

Hermanson, J. W., and T. J. O'Shea. 1983. *Antrozous pallidus. Mammalian Species* 213: 1–8.

Herring, P. J. 2007. Sex with the lights on? A review of bioluminescent sexual dimorphism in the sea. *Journal of the Marine Biological Association of the United Kingdom* 87: 829–42. doi:10.1017/S0025315407056433.

Herzner, G., A. Schlecht, V. Dollhofer, C. Parzefall, K. Harrar, A. Kreuzer, L. Pilsl, and J. Ruther. 2013. Larvae of the parasitoid wasp *Ampulex compressa* sanitize their host, the American cockroach, with a blend of antimicrobials. *Proceedings of the National Academy of Sciences USA* 110: 1369–74. doi:10.1073/pnas.1213384110.

Hill, G. E. 2000. Energetic constraints on expression of carotenoid-based plumage coloration. *Journal of Avian Biology* 31: 559–66. doi:10.1034/j.1600-048X.2000.310415.x.

Holstein, T., and P. Tardent. 1984. An ultrahigh-speed analysis of exocytosis: Nematocyst discharge. *Science* 223: 830–33. doi:10.1126/science.6695186.

House, P. K., A. Vyas, and R. Sapolsky. 2011. Predator cat odors activate sexual arousal pathways in brains of *Toxoplasma gondii* infected rats. *PLoS One* 6: e23277. doi:10.1371/journal.pone.0023277.

Hoving, H. J. T., S. L. Bush, and B. H. Robinson. 2011. A shot in the dark: Same-sex sexual behaviour in a deep-sea squid. *Biology Letters* 8: 287–90. doi:10.1098/rsbl.2011.0680.

Hudson, P. J., A. P. Dobson, and K. D. Lafferty. 2006. Is a healthy ecosystem one that is rich in parasites? *Trends in Ecology and Evolution* 21: 381–85. doi:10.1016/j.tree.2006.04.007.

Ignarsson, I., M. Kuntner, and L. J. May-Collado. 2010. Dogs, cats, and kin: a molecular species-level phylogeny of Carnivora. *Molecular Phylogenetics and Evolution* 54: 726–45. doi:10.1016/j.ympev.2009.10.033.

Isbister, G. K., and H. W. Fan. 2011. Spider bite. *Lancet* 378: 2039–47. doi:10.1016/S0140-6736(10)62230-1.

Isbister, G. K., M. R. Gray, C. R. Balit, R. J. Raven, B. J. Stokes, K. Porges, A. S. Tankel, E. Turner, J. White, and M. M. Fisher. 2005. Funnel-web spider bite: A systematic review of recorded clinical cases. *Medical Journal of Australia* 182: 407–11.

Iyengar, E. V. 2008. Kleptoparasitic interactions throughout the animal kingdom and a re-evaluation, based on participant mobility, of the conditions promoting the evolution of kleptoparasitism. *Biological Journal of the Linnean Society* 93: 745–62. doi:10.1111/j.1095-8312.2008.00954.x.

Izzo, T. J., D. J. Rodrigues, M. Menin, A. P. Lima, and W. E. Magnusson. 2012. Functional necrophilia: A profitable anuran reproductive strategy? *Journal of Natural History* 46: 2961–67. doi:10.1080/00222933.2012.724720.

Jablonski, D. 1994. Extinctions in the fossil record. *Philosophical Transactions of the Royal Society of London Series B: Biological Sciences* 344: 11–16. doi:10.1098/rstb.1994.0045.

Janzen, D. H., and P. S. Martin. 1982. Neotropical anachronisms: The fruits the gomphotheres ate. *Science* 215: 19–27. doi:10.1126/science.215.4528.19.

Karban, R., M. Huntzinger, and A. C. McCall. 2004. The specificity of eavesdropping on sagebrush by other plants. *Ecology* 85: 1846–52. doi:10.1890/03-0593.

King, A. J., A. M. Wilson, S. D. Wilshin, J. Lowe, H. Haddadi, S. Hailes, and A. J. Morton. 2012. Selfish-herd behaviour of sheep under threat. *Current Biology* 22: R561–62. doi:10.1016/j.cub.2012.05.008.

Kjellesvig-Waering, E. N. 1972. *Brontoscorpio anglicus*: A gigantic lower Paleozoic scorpion from central England. *Journal of Paleontology* 46: 39–42.

Koch, C. A., S. M. Olsen, and E. J. Moore. 2012. Use of the medicinal leech for salvage of venous congested microvascular free flaps of the head and neck. *American Journal of Otolaryngology—Head and Neck Medicine and Surgery* 33: 26–30. doi:10.1016/j.amjoto.2010.12.004.

Kushnir, H., H. Leitner, D. Ikanda, and C. Packer. 2010. Human and ecological risk factors for unprovoked lion attacks on humans in Southeastern Tanzania. *Human Dimensions of Wildlife* 15: 315–31. doi:10.1080/10871200903510999.

Lafferty, K. D. 2006. Can the common brain parasite, *Toxoplasma gondii*, influence human culture? *Proceedings of the Royal Society B* 273: 2749–55. doi:10.1098/rspb.2006.3641.

Laws, R. M. 1970. Elephants as agents of habitat and landscape change in East Africa. *Oikos* 21: 1–15.

Leitner, M., W. Boland, and A. Mithöfer. 2005. Direct and indirect defenses induced by piercing-sucking and chewing herbivores in *Medicago trunculata*. *New Phytologist* 167: 597–606. doi:10.1111/j.1469-8137.2005.01426.x.

Lewis, R. J., S. Dutertre, I. Vetter, and M. J. Christie. 2012. *Conus* venom peptide pharmacology. *Pharmacological Reviews* 64: 259–98. doi:10.1124/pr.111.005322.

Li, D., J. Oh, S. Kralj-Fišer, and M. Kuntner. 2012. Remote copulation: male adaptation to female cannibalism. *Biology Letters* 8: 512–15. doi:10.1098/rsbl.2011.1202.

Liebert, P. S., and R. C. Madden. 2004. Human botfly larva in a child's scalp. *Journal of Pediatric Surgery* 39: 629–30. doi:10.1016/j.jpedsurg.2003.12.035.

Lockyer, C. 1976. Body weights of some species of large whales. *Journal du Conseil International pour l'Exploration de la Mer* 36: 259–73. doi:10.1093/icesjms/36.3.259.

Maity, P. and S. A. Tekalur. 2011. Finite element analysis of ramming in *Ovis canadensis*. *Journal of Biomechanical Engineering* 133: 021009. doi:10.1115/1.4003321.

Martin, A. M., H. Presseault-Gauvin, M. Festa-Bianchet, and F. Pelletier. 2013. Male mating competitiveness and age-dependent relationship between testosterone and social rank in bighorn sheep. *Behavioral Ecology and Sociobiology* 67: 919–28.doi:10.1007/s00265-013-1516-7.

McKinney, F., and S. Evarts. 1997. Sexual coercion in waterfowl and other birds. *Ornithological Monographs* 49: 163–95. doi:10.2307/40166723.

Meehan, C. J., E. J. Olson, M. W. Reudink, T. K. Kyser, and R. L. Curry. 2009. Herbivory in a spider through exploitation of an ant–plant mutualism. *Current Biology* 19: R892–93. doi:10.1016/j.cub.2009.08.049.

Meredith, R. W., J. E. Janečka, J. Gatesy, O. A. Ryder, C. A. Fisher, E. C. Teeling, A. Goodbla, et al. 2011. Impacts of the Cretaceous terrestrial revolution and KPg extinction on mammal diversification. *Science* 334: 521–24. doi:10.1126/science.1211028.

Meyer, R. S., A. E. DuVal, and H. R. Jensen. 2012. Patterns and processes in crop domestication: An historical review and quantitative analysis of 203 global food crops. *New Phytologist* 196: 29–48. doi:10.1111/j.1469-8137.2012.04253.x.

Michiels, N. K., and L. J. Newman. 1998. Sex and violence in hermaphrodites. *Nature* 391: 647. doi:10.1038/35527.

Miller, G. H., M. L. Fogel, J. W. Magee, M. K. Gagan, S. J. Clarke, and B. J. Johnson. 2005. Ecosystem collapse in Pleistocene Australia and a human role in megafaunal extinction. *Science* 309: 287–90. doi:10.1126/science.1111288.

Mills, M. G. L., and T. M. Shenk. 1992. Predator-prey relationships: The impact of lion predation on wildebeest and zebra populations. *Journal of Animal Ecology* 61: 693–702.

Min, K.-J., C.-K. Lee, and H.-N. Park. 2012. The lifespan of Korean eunuchs. *Current Biology* 22: R792–93. doi:10.1016/j.cub.2012.06.036.

Mitchell, G. C., and J. R. Tigner. 1970. The route of ingested blood in the vampire bat (*Desmodus rotundus*). *Journal of Mammalogy* 51: 814–17.

Mithöfer, A., and W. Boland. 2012. Plant defense against herbivores: Chemical aspects. *Annual Review of Plant Biology* 63: 431–50. doi:10.1146/annurev-arplant-042110-103854.

Mock, D. W., H. Drummond, and C. H. Stinson. 1990. Avian siblicide. *American Scientist* 78: 438–49.

Montoya, J. G., and O. Liesenfeld. 2004. Toxoplasmosis. *Lancet* 363: 1965–76. doi:10.1016/S0140-6736(04)16412-X.

Morrison, P. R., M. Pierce, and F. A. Ryser. 1957. Food consumption and body weight in the masked and short-tail shrews. *American Midland Naturalist* 57: 493–501.

Muchhala, N. 2006. Nectar bat stows huge tongue in its rib cage. *Nature* 444: 701. doi:10.1038/444701a.

Muchhala, N., and J. D. Thomson. 2009. Going to great lengths: Selection for long corolla tubes in an extremely specialized bat-flower mutualism. *Proceedings of the Royal Society of London B* 276: 2147–52. doi:10.1098/rspb.2009.0102.

Murie, O. J. 1929. Nesting of the snowy owl. *Condor* 31: 3–12.

Naylor, R., S. J. Richardson, and B. M. McAllan. 2008. Boom and bust: A review of the physiology of the marsupial genus *Antechinus*. *Journal of Comparative Physiology B* 178: 545–62. doi:10.1007/s00360-007-0250-8.

Neaves, W. B., and P. Baumann. 2011. Unisexual reproduction among vertebrates. *Trends in Genetics* 27: 81–88. doi:10.1016/j.tig.2010.12.002.

Nelson, L. 2004. Venomous snails: One slip, and you're dead. . . . *Nature* 429: 798–99. doi:10.1038/429798a.

Nicholson, G. M., A. Graudins, H. I. Wilson, M. Little, and K. W. Broady. 2006. Arachnid toxinology in Australia: From clinical toxicology to potential applications. *Toxicon* 48: 872–98. doi:10.1016/j.toxicon.2006.07.025.

O'Connell, J. F., K. Hawkes, and N. Blurton Jones. 1988. Hadza scavenging: Implications for Plio/Pleistocene hominid subsistence. *Current Anthropology* 29: 356–63.

Packer, C. 2000. Infanticide is no fallacy. *American Anthropologist* 102: 829–31.

Packer, C., D. Ikanda, B. Kissui, and H. Kushnir. 2005. Lion attacks on humans in Tanzania. *Nature* 436: 927–28. doi:10.1038/436927a.

Page, R. A., and M. J. Ryan. 2005. Flexibility in assessment of prey cues: Frog-eating bats and frog calls. *Proceedings of the Royal Society B* 272: 841–47. doi:10.1098/rspb.2004.2998.

Paré, P. W., and J. H. Tumlinson. 1999. Plant volatiles as a defense against insect herbivores. *Plant Physiology* 121: 325–31. doi:10.1104/pp.121.2.325.

Parmelee, D. F. 1992. Snowy Owl (*Bubo scandiacus*). In *The Birds of North America Online*, edited by A. Poole. Ithaca, NY: Cornell Lab of Ornithology. http://bna.birds.cornell.edu/bna/species/010.

Parvinen, K., and U. Dieckmann. 2013. Self-extinction through optimizing selection. *Journal of Theoretical Biology* 333: 1–9. doi:10.1016/j.jtbi.2013.03.025.

Pasitschniak-Arts, M., and S. Larivière. 1995. *Gulo gulo*. *Mammalian Species* 499: 1–10.

Pimentel, D., and A. Wilson. 2010. Non-indigenous species: Crops and livestock. In *Encyclopedia of Pest Management,* Vol. II, edited by D. Pimentel, 400–403. Boca Raton, FL: CRC Press.

Platnick, N. I. 2013. *The World Spider Catalog, Version 14.0*. New York: American Museum of Natural History, online at http://research.amnh.org/entomology/spiders/catalog/index.html. doi:10.5531/db.iz.0001.

Pluháček, J., and L. Bardoš. 2000. Male infanticide in captive plains zebra, *Equus burchelli*. *Animal Behaviour* 59: 689–94. doi:10.1006/anbe.1999.1371.

Polis, G. A. 1990. Introduction. In *The Biology of Scorpions*, edited by G. A. Polis, 1–8. Stanford, CA: Stanford University Press.

Poulin, R., and S. Morand. 2000. The diversity of parasites. *Quarterly Review of Biology* 75: 277–93.

Prescott, G. W., D. R. Williams, A. Balmford, R. E. Green, and A. Manica. 2012. Quantitative global analysis of the role of climate and people in explaining late Quaternary megafaunal extinctions. *Proceedings of the National Academy of Sciences USA* 109: 4527–31 doi:10.1073/pnas.1113875109.

Proctor, H. C. 2003. Feather mites (Acari: Astigmata): Ecology, behavior, and evolution. *Annual Review of Entomology* 48: 185–209. doi:10.1146/annurev.ento.48.091801.112725.

Range, F., L. Horn, Z. Viranyi, and L. Huber. 2009. The absence of reward induces inequity aversion in dogs. *Proceedings of the National Academy of Sciences USA* 106: 340–45. doi:10.1073/pnas.0810957105.

Ranger, C. M., R. E. Winter, A. P. Singh, M. E. Reding, J. M. Frantz, J. C. Locke, and C. R. Krause. 2011. Rare excitatory amino acid from flowers of zonal geranium responsible for paralyzing the Japanese beetle. *Proceedings of the National Academy of Sciences USA* 108: 1217–21. doi:10.1073/pnas.1013497108.

Ratcliffe, J. M., M. B. Fenton, and B. G. Galef Jr. 2003. An exception to the rule: Common vampire bats do not learn taste aversions. *Animal Behaviour* 65: 385–89. doi:10.1006/anbe.2003.2059.

Reinhardt, K., and M. T. Siva-Jothy. 2007. Biology of the bed bugs (Cimicidae). *Annual Review of Entomology* 52: 351–74. doi:10.1146/annurev.ento.52.040306.133913.

Resendes, A. R., A. F. S. Amaral, A. Rodrigues, and S. Almeria. 2009. Prevalence of *Calodium hepaticum* (Syn. *Capillaria hepatica*) in house mice (*Mus musculus*) in the Azores archipelago. *Veterinary Parasitology* 160: 340–43. doi:10.1016/j.vetpar.2008.11.001.

Riskin, D. K., and M. B. Fenton. 2001. Sticking ability in Spix's disk-winged bat, *Thyroptera tricolor* (Microchiroptera: Thyropteridae). *Canadian Journal of Zoology* 79: 2261–67. doi:10.1139/z01-192.

Riskin, D. K., and P. A. Racey. 2010. How do sucker-footed bats hold on, and why do they roost head-up? *Biological Journal of the Linnean Society* 99: 233–40. doi:10.1111/j.1095-8312.2009.01362.x.

Riskin, S. H., S. Porder, C. Neill, A. M. E. S. F. Figueira, C. Tubbesing, and N. Mahowald. 2013. The fate of phosphorus fertilizer in Amazon soya bean fields. *Philosophical Transactions of the Royal Society B* 368: 20120154. doi:10.1098/rstb.2012.0154.

Rivers, T. J., and J. G. Morin. 2009. Plasticity of male mating behaviour in a marine bioluminescent ostracod in both time and space. *Animal Behaviour* 78: 723–34. doi:10.1016/j.anbehav.2009.06.020.

Rothman, J. M., E. S. Dierenfeld, D. O. Molina, A. V. Shaw, H. F. Hintz, and A. N. Pell. 2006. Nutritional chemistry of foods eaten by gorillas in Bwindi Impenetrable National Park, Uganda. *American Journal of Primatology* 68: 675–91. doi:10.1002/ajp.20243.

Rowe, F. P., E. J. Taylor, and A. H. J. Chudley. 1964. The effect of crowding on the reproduction of the house-mouse (*Mus musculus* L.) living in corn-ricks. *Journal of Animal Ecology* 33: 477–83.

Rumpho, M. E., J. M. Worful, J. Lee, K. Kannan, M. S. Tyler, D. Bhattacharya, A. Moustafa, and J. R. Manhart. 2008. Horizontal gene transfer of the algal nuclear gene *psbO* to the photosynthetic sea slug *Elysia chlorotica*. *Proceedings of the National Academy of Sciences USA* 105: 17867–71. doi:10.1073/pnas.0804968105.

Schmidt, J. O. 1990. Hymenopteran venoms: Striving toward the ultimate defense against vertebrates. In *Insect defenses: Adaptive mechanisms and strategies of prey and predators*, edited by D. L. Evans and J. O. Schmidt, 387–419. Albany, NY: State University of New York Press.

Schmidt, J. O., M. S. Blum, and W. L. Overal. 1983. Hemolytic activities of stinging insect venoms. *Archives of Insect Biochemistry and Physiology* 1: 155–60. doi:10.1002/arch.940010205.

Schnell, I. B., P. F. Thomsen, N. Wilinson, M. Rasmussen, L. R. D. Jensen, E. Willerslev, M. F. Bertelsen, and M. T. P. Gilbert. 2012. Screening mammal biodiversity using DNA from leeches. *Current Biology* 22: R262–63. doi:10.1016/j.cub.2012.02.058.

Scholes, E. 2008. Evolution of the courtship phenotype in the bird of paradise genus *Parotia* (Aves: Paradisaeidae): Homology, phylogeny, and modularity. *Biological Journal of the Linnean Society* 94: 491–504. doi:10.1111/j.1095-8312.2008.01012.x:.

Schutt, W. A. Jr., J. S. Altenbach, Y. H. Chang, D. M. Cullinane, J. W. Hermanson, F. Muradali, and J. E. A. Bertram. 1997. Dynamics of flight-initiating jumps in the common vampire bat *Desmodus rotundus*. *Journal of Experimental Biology* 200: 3003–12.

Sessions, A. L., D. M. Doughty, P. V. Welander, R. E. Summons, and D. K. Newman. 2009. The continuing puzzle of the Great Oxidation Event. *Current Biology* 19: R567–74. doi:10.1016/j.cub.2009.05.054.

Shine, R., T. Langkilde, and R. T. Mason. 2003. Cryptic forcible insemination: Male snakes exploit female physiology, anatomy, and behavior to obtain coercive matings. *The American Naturalist* 162: 653–67. doi:10.1086/378749.

Showstack, R. 2013. Largest meteor since Tunguska event explodes above Russian city. *Eos: Transactions, American Geophysical Union* 94: 87. doi:10.1002/2013EO090004.

Shrestha, L. B. 2006. Life expectancy in the United States. Congressional Research Service Report for Congress.

Sidor, C. A., D. A. Vilhena, K. D. Angielczyk, A. K. Huttenlocker, S. J. Nesbitt, B. R. Peecook, J. S. Steyer, R. M. H. Smith, and L. A. Tsuji. 2013. Provincialization of terrestrial faunas following the end-Permian mass extinction. *Proceedings of the National Academy of Sciences USA* 110: 8129–33. doi:10.1073/pnas.1302323110.

Silber, G. K., M. W. Newcomer, and H. M. Pérez-Cortés. 1990. Killer whales (*Orcinus orca*) attack and kill a Bryde's whale (*Balaenoptera edeni*). *Canadian Journal of Zoology* 68: 1603–6. doi:10.1139/z90-238.

Sorenson, L. G., and S. R. Derrickson. 1994. Sexual selection in the northern pintail (*Anas acuta*): The importance of female choice versus male-male competition in the evolution of sexually-selected traits. *Behavioral Ecology and Sociobiology* 35: 389–400. doi:10.1007/BF00165841.

Stålhandske, P. 2001. Nuptial gift in the spider *Pisaura mirabilis* maintained by sexual selection. *Behavioral Ecology* 12: 691–97. doi:10.1093/beheco/12.6.691.

Steudte, S., I. Kolassa, T. Stalder, A. Pfeiffer, C. Kirschbaum, and T. Elbert. 2011. Increased cortisol levels in hair of severely traumatized Ugandan individuals with PTSD. *Psychoneuroendocrinology* 36: 1193–200. doi:10.1016/j.psyneuen.2011.02.012.

Sullivan, B. K. 1983. Sexual selection in the Great Plains toad (*Bufo cognatus*). *Behaviour* 84: 258–64.

Šuput, D. 2009. In vivo effects of cnidarian toxins and venoms. *Toxicon* 54: 1190–200. doi:10.1016/j.toxicon.2009.03.001.

Swinburn, B. A., G. Sacks, K. D. Hall, K. McPherson, D. T. Finegood, M. L. Moodie, and S. L. Gortmaker. 2011. The global obesity pandemic: Shaped by global drivers and local environments. *Lancet* 378: 804–14. doi:10.1016/S0140-6736(11)60813-1.

Szykman, M., R. C. Van Horn, A. L. Engh, E. E. Boydston, and K. E. Holekamp. 2007. Courtship and mating in free-living spotted hyenas. *Behaviour* 144: 815–46. doi:10.1163/156853907781476418.

Tellgren-Roth, Å., K. Dittmar, S. E. Massey, C. Kemi, C. Tellgren-Roth, P. Savolainen, L. A. Lyons, and D. A. Liberles. 2009. Keeping the blood flowing: Plasminogen activator genes and feeding behavior in vampire bats. *Naturwissenschaften* 96: 39–47. doi:10.1007/s00114-008-0446-0.

Thomas, R. G., and F. H. Pough. 1979. The effect of rattlesnake venom on digestion of prey. *Toxicon* 17: 221–28. doi:10.1016/0041-0101(79)90211-3.

Thompson, T. E., and I. Bennett. 1969. *Physalia* nematocysts: Utilized by mollusks for defense. *Science* 166: 1532–33. doi:10.1126/science.166.3912.1532.

Topoff, H. 1999. Slave-making queens. *Scientific American* 281: 84–90.

Tranah, G. J., T. M. Manini, K. K. Lohman, M. A. Nalls, S. Kritchevsky, A. B. Newman, T. B. Harris, I. Miljovich, A. Biffi, S. R. Cummings, and Y. Liu. 2011. Mitochondrial DNA variation in human metabolic rate and energy expenditure. *Mitochondrion* 11: 855–61. doi:10.1016/j.mito.2011.04.005.

Treat, A. E. 1957. Unilaterality in infestations of the moth ear mite. *Journal of the New York Entomological Society* 65: 41–50.

Uetz, G. W., A. McCrate, and C. S. Hieber. 2010. Stealing for love? Apparent nuptial gift behavior in a kleptoparasitic spider. *Journal of Arachnology* 38: 128–31. doi:10.1636/Hi08-100.1.

Urrutia-Fucugauchi, J., A. Camargo-Zanoguera, and L. Pérez-Cruz. 2011. Discovery and focused study of the Chicxulub impact crater. *Eos: Transactions, American Geophysical Union* 92: 209–10. doi:10.1029/2011EO250001.

Utami, S. C., and J. A. R. A. M. Van Hooff. 1997. Meat-eating by adult female Sumatran orangutans (*Pongo pygmaeus abelii*). *American Journal of Primatology* 43: 159–65. doi:10.1002/(SICI)1098-2345(1997)43:2<159::AID-AJP5>3.0.CO;2-W.

Valmalette, J. C., A. Dombrovsky, P. Brat, C. Mertz, M. Capovilla, and A. Robichon. 2012. Light-induced electron transfer and ATP synthesis in a carotene synthesizing insect. *Scientific Reports* 2: 579. doi:10.1038/srep00579.

van der Meer, E., M. Moyo, G. S. A. Rasmussen, and H. Fritz. 2011. An empirical and experimental test of risk and costs of kleptoparasitism for African wild dogs (*Lycaon pictus*) inside and outside a protected area. *Behavioral Ecology* 22: 985–92. doi:10.1093/beheco/arr079.

van Wolkenten, M., S. F. Brosnan, and F. B. M. de Waal. 2007. Inequity responses of monkeys modified by effort. *Proceedings of the National Academy of Sciences USA* 104: 18854–18859. doi:10.1073/pnas.0707182104.

Vass, A. A. 2001. Beyond the grave: Understanding human decomposition. *Microbiology Today* 28: 190–92.

Verbrugge, L. M., J. J. Rainey, R. L. Reimnick, and H. D. Blankespoor. 2004. Swimmer's itch: Incidence and risk factors. *American Journal of Public Health* 94: 738–41.

Vetter, J. 2000. Plant cyanogenic glycosides. *Toxicon* 38: 11–36. doi:10.1016/S0041-0101(99)00128-2.

Vetter, R. S., and G. K. Isbister. 2008. Medical aspects of spider bites. *Annual Review of Entomology* 53: 409–29. doi:10.1146/annurev.ento.53.103106.093503.

Vieites, D. R., S. Nieto-Román, M. Barluenga, A. Palanca, M. Vences, and A. Meyer. 2004. Post-mating clutch piracy in an amphibian. *Nature* 431: 305–308. doi:10.1038/nature02879.

Voigt, C. C., O. Behr, B. Caspers, O. von Helversen, M. Knörnschild, F. Mayer, and M. Nagy. 2008. Songs, scents, and senses: Sexual selection in the greater sac-winged bat, *Saccopteryx bilineata*. *Journal of Mammalogy* 89: 1401–10. doi:10.1644/08-MAMM-S-060.1.

Voigt, C. C., B. Caspers, and S. Speck. 2005. Bats, bacteria, and bat smell: Sex-specific diversity of microbes in a sexually selected scent organ. *Journal of Mammalogy* 86: 745–49.

Voigt, C. C., and O. von Helversen. 1999. Storage and display of odour by male *Saccopteryx bilineata* (Chiroptera, Emballonuridae). *Behavioral Ecology and Sociobiology* 47: 29–40. doi:10.1007/s002650050646.

Vollrath, F. 1978. A close relationship between two spiders (Arachnida, Araneidae): *Curimagua bayano* synecious on a *Diplura* species. *Psyche* 85: 347–54. doi:10.1155/1978/27439.

——. 1979. Vibrations: Their signal function for a spider kleptoparasite. *Science* 205: 1149–51. doi:10.1126/science.205.4411.1149.

——. 1998. Dwarf males. *Trends in Ecology and Evolution* 13: 159–63. doi:10.1016/S0169-5347(97)01283-4.

Vyas, A., S. Kim, N. Giacomini, J. C. Boothroyd, and R. M. Sapolsky. 2007. Behavioral changes induced by *Toxoplasma* infection of rodents are highly specific to aversion of cat odors. *Proceedings of the National Academy of Sciences USA* 104: 6442–47. doi:10.1073/pnas.0608310104.

Wanless, R. M., A. Angel, R. J. Cuthbert, G. M. Hilton, and P. G. Ryan. 2007. Can predation by invasive mice drive seabird extinctions? *Biology Letters* 3: 241–44. doi:10.1098/rsbl.2007.0120.

Warrell, D. A. 2010. Snake bite. *Lancet* 375: 77–88. doi:10.1016/S0140-6736(09)61754-2.

Waters, A., F. Blanchette, and A. D. Kim. 2012. Modeling huddling penguins. *PLoS One* 7: e50277. doi:10.1371/journal.pone.0050277.

Watts, H. E., J. B. Tanner, B. L. Lundrigan, and K. E. Holekamp. 2009. Post-weaning maternal effects and the evolution of female dominance in the spotted hyena. *Proceedings of the Royal Society B* 276: 2291–98. doi:10.1098/rspb.2009.0268.

Way, J. L. 1984. Cyanide intoxication and its mechanism of antagonism. *Annual Reviews of Pharmacology and Toxicology* 24: 451–81.

Welch, K. D., K. E. Panter, S. T. Lee, D. R. Gardner, B. L. Stegelmeier, and D. Cook. 2009. Cyclopamine-induced synophthalmia in sheep: defining a critical window and toxicokinetic evaluation. *Journal of Applied Toxicology* 29: 414–21. doi:10.1002/jat.1427.

Wikelski, M., and S. Bäurle. 1996. Pre-copulatory ejaculation solves time constraints during copulations in marine iguanas. *Proceedings of the Royal Society B: Biological Sciences* 263: 439–44. doi:10.1098/rspb.1996.0066.

Wilkinson, G. S. 1984. Reciprocal food sharing in the vampire bat. *Nature* 308: 181–84. doi:10.1038/308181a0.

Windmill, J. F. C., J. H. Fullard, and D. Robert. 2007. Mechanics of a "simple" ear: Tympanal vibrations in noctuid moths. *Journal of Experimental Biology* 210: 2637–48. doi:10.1242/jeb.005025.

Wojcieszek, J. M., J. A. Nicholls, and A. W. Goldizen. 2007. Stealing behavior and the maintenance of a visual display in the satin bowerbird. *Behavioral Ecology* 18: 689–95. doi:10.1093/beheco/arm031.

Wong, S., and J. S. Remington. 1994. Toxoplasmosis in pregnancy. *Clinical Infectious Diseases* 18: 853–61.

World Health Organization. 2005. *World Health Report 2005: Make Every Mother and Child Count*. http://www.who.int/whr/2005/en/index.html.

——. 2012. *Fact Sheet Number 348: Maternal Mortality.* Updated May 2012. http://www.who.int/mediacentre/factsheets/fs348/en/index.html.

Yanagihara, A. A., and R. V. Shohet. 2012. Cubozoan venom-induced cardiovascular collapse is caused by hyperkalemia and prevented by zinc gluconate in mice. *PLoS One* 7: e51368. doi:10.1371/journal.pone.0051368.

Yosef, R., and B. Pinshow. 2005. Impaling in true shrikes (Laniidae): A behavioral and ontogenetic perspective. *Behavioural Processes* 69: 363–67. doi:10.1016/j.beproc.2005.02.023.

Zhang, L., X. Yang, H. Wu, X. Gu, Y. Hu, and F. Wei. 2011. The parasites of giant pandas: Individual-based measurement in wild animals. *Journal of Wildlife Diseases* 47: 164–71.

Zimmer, C. 2000. *Parasite Rex*. New York: Free Press.

Zimmermann, T. 2010. The killer in the pool. *Outside Magazine.* Published online July 30, 2010. http://www.outsideonline.com/outdoor-adventure/nature/The-Killer-in-the-Pool.html?page=all.

INDEX

Index

Index

Index

Index

Index

ABOUT THE AUTHOR

Dan Riskin was born in 1975 in Edmonton, Canada. He did his undergraduate work at the University of Alberta (BSc, 1997), his master's at York University (MSc, 2000), and his doctorate at Cornell University (PhD, 2006). He did postdoctoral research at Brown University (2006–10) and at the Center for Ecology and Conservation Biology of Boston University (2006–07). His work is mostly focused on the biomechanics of bat locomotion—the physics behind crawling on the ground, jumping into flight, flapping through the sky, and landing on the ceiling. Dan has traveled around the world for his research and won numerous awards for his research and teaching. He is the author or coauthor of more than twenty refereed articles in such journals as the *Journal of Experimental Biology*, the *Biological Journal of the Linnean Society*, and *Nature*.

Dan's first work for TV was on the Emmy-nominated History Channel show *Evolve*. Soon afterward, he contributed to Animal Planet's *Monsters Inside Me*, and because of the success of that show, he appeared on *The Dr. Oz Show*, *The Tonight Show with Jay Leno*, and *The Late Late Show with Craig Ferguson*. Dan has also appeared on *Human Nature* (Discovery Science) and *Bedbug Apocalypse* (Animal Planet).

In 2011, Dan left academics to become the cohost of the world's only daily science program, *Daily Planet*, on Discovery Canada. He is a regular guest on *Canada AM*, *CTV NewsChannel*, and *CTV National News* in Canada, and in the USA on *The Late Late Show with Craig Ferguson*.

Dan lives in Toronto with his wife, Shelby; their son, Sam; and their Boston terrier, Elliott. He is on Twitter as @riskindan. This is his first book.